International Workshop on Positron Studies of Defects 2014

Journal of Physics: Conference Series Volume 674

Kyoto, Japan
14 – 19 September 2014

Editors:

Kazuki Sugita
Yasuharu Shirari

ISBN: 978-1-5108-1965-8
ISSN: 1742-6588

Printed from e-media with permission by:

Curran Associates, Inc.
57 Morehouse Lane
Red Hook, NY 12571

Some format issues inherent in the e-media version may also appear in this print version.

Printed by Curran Associates, Inc. (2016)

For permission requests, please contact the Institute of Physics
at the address below.

Institute of Physics
Dirac House, Temple Back
Bristol BS1 6BE UK

Phone: 44 1 17 929 7481
Fax: 44 1 17 920 0979

techtracking@iop.org

Additional copies of this publication are available from:

Curran Associates, Inc.
57 Morehouse Lane
Red Hook, NY 12571 USA
Phone: 845-758-0400
Fax: 845-758-2634
Email: curran@proceedings.com
Web: www.proceedings.com

Table of contents

Volume 674

The International Workshop on Positron Studies of Defects 2014
14–19 September 2014, Kyoto, Japan

Accepted papers received: 1 December 2015
Published online: 26 January 2016

Preface

Papers

Defects in metals

Defects in metal oxides

Polymers and soft matter

Theoretical calculations

Positron beam studies

Journal of Physics: Conference Series **674** (2016) 011001 doi:10.1088/1742-6596/674/1/011001

Abstract

The International Workshop on Positron Studies of Defects 2014 (PSD-14) was held in Kyoto, Japan from 14–19 September, 2014. The PSD Workshop brought together positron scientists interested in studying defects to an international platform for presenting and discussing recent results and achievements, including new experimental and theoretical methods in the field. The workshop topics can be characterized as follows:

- Positron studies of defects in semiconductors and oxides
- Positron studies of defects in metals
- New experimental methods and equipment
- Theoretical calculations and simulations of momentum distributions, positron lifetimes and other characteristics for defects
- Positron studies of defects in combination with complementary methods
- Positron beam studies of defects at surfaces, interfaces, in sub-surface regions and thin films
- Nanostructures and amorphous materials

The first PSD workshop was organized in Wernigerode, Germany in 1987. The next four workshops were held in Halle, Germany (1994), Hamilton, Canada (1999), Sendai, Japan (2002) and Pullman, USA (2004) under the name Positron Studies of Semiconductor Defects (PSSD). As studying defects in metals is once again gaining importance – as is also documented in these Proceedings – the name of the workshop in Prague was changed to the original version PSD. Then the PSD workshops were held in Prague, Czech Republic (2008), Delft, Netherlands (2011) and Kyoto, Japan (2014). The next workshop (PSD-17) will be held in Dresden-Rossendorf, Germany in September 2017.

We would like to express our deep gratitude to all of the workshop participants for their presentations and contributions to discussions, which made the PSD workshop a successful scientific meeting. In total, 105 scientists and students from 15 countries took part in the PSD workshop. The workshop programme was comprised of 13 invited lectures, 30 contributed talks and 36 poster presentations. Twenty-five of the contributions are included in this proceedings, covering various aspects of defect studies by using positrons. The proceedings papers are divided into five categories: defects in metals, defects in metal oxides, polymers and soft matter, theoretical calculations, and positron beam studies. We hope new results and ideas in these proceedings will be very

Journal of Physics: Conference Series **674** (2016) 011001 doi:10.1088/1742-6596/674/1/011001

useful in improving our understanding of positron studies of defects.

Finally, on behalf of all the members of our Organizing Committee and International Advisory Committee, we wish to express our sincere gratitude to all of the participants in PSD-14 and the authors of this proceedings. We look forward to seeing everyone in Germany in 2017.

Kazuki Sugita,

Yasuharu Shirai

Editors

Local Organizing Committee

H. Araki (Osaka Univ.)
M. Fujinami (Chiba Univ.)
F. Hori (Osaka Pref. Univ.)
K. Inoue (Tohoku Univ.)
A. Kinomura (Kyoto Univ.)
M. Mizuno (Osaka Univ.)
Y. Nagai (Tohoku Univ.)
K. Sakaki (AIST)
Y. Shirai (Kyoto Univ.) -Chair
K. Sugita (Kyoto Univ.)

International Advisory Committee

J. Cizek (Charles University, Prague, Czech Republic)
Chr. Hugenschmidt (TU Munich, Germany)
R. Krause-Rehberg (Martin-Luther University Halle, Germany)
K.G. Lynn (Washington State University, Pullman, USA)
P. Mascher (McMaster University, Canada)
H. Schut (Delft University, The Netherlands)
Y. Shirai (Kyoto University, Japan)
F. Tuomisto (Aalto University, Helsinki, Finland)

PSD 14

Journal of Physics: Conference Series **674** (2016) 011002 doi:10.1088/1742-6596/674/1/011002

Peer review statement

All papers published in this volume of *Journal of Physics: Conference Series* have been peer reviewed through processes administered by the proceedings Editors. Reviews were conducted by expert referees to the professional and scientific standards expected of a proceedings journal published by IOP Publishing.

Journal of Physics: Conference Series **674** (2016) 012001 doi:10.1088/1742-6596/674/1/012001

Spatial distribution of defects in ultra fine grained copper prepared by high pressure torsion

J Čížek[1], O Melikhova[1], I Procházka[1], M Janeček[2], P Hruška[1] and S Dobatkin[3]

[1]Charles University in Prague, Faculty of Mathematics and Physics, V Holešovičkách 2, 18000 Praha 8, Czech Republic
[2]Charles University in Prague, Faculty of Mathematics and Physics, Ke Karlovu 3, 12116 Praha 2, Czech Republic
[3]A.A. Baikov Institute of Metallurgy and Materials Science, Russian Academy of Sciences, Moscow, Russia

E-mail: jakub.cizek@mff.cuni.cz

Abstract. Bulk materials with ultra fine grain structure can be fabricated by severe plastic deformation. Among variety of techniques based on severe plastic deformation high pressure torsion is the most efficient method for grain refinement down to nano-scale. In torsion deformation the strain distribution across the sample is non-uniform and increases with increasing radial distance from the centre of the sample corresponding to the axis of torsional straining. Due to this reason it is very important to examine homogeneity of ultra fine grained structure of samples prepared by high pressure torsion. In the present work positron annihilation spectroscopy was employed for mapping of spatial distribution of defects in ultra fine grained copper prepared by high pressure torsion. Spatial distribution of defects was examined by means of (i) Doppler broadening using S parameter for mapping of defect density and (ii) positron lifetime spectroscopy. Spatially resolved positron annihilation studies were combined with mapping by microhardness testing. Hardness is sensitive to dislocation density due to work hardening but is practically not affected by vacancies while positron annihilation is sensitive both to dislocations and vacancies. Our investigations revealed that ultra fine grained copper contains dislocations and vacancy clusters created by agglomeration of deformation-induced vacancies. Average size of vacancy clusters increases with increasing radial distance from the centre of the sample due to higher production rate of vacancies resulting in larger clusters. During high pressure torsion deformation microhardness increases firstly at the periphery of the sample due to the highest imposed strain. With increasing number of high pressure torsion revolutions the hardness increases also in the centre and finally becomes practically uniform across the whole sample indicating the homogeneous distribution of dislocations. Doppler broadening mapping revealed a remarkable increase of S parameter at the sample periphery due to larger size of vacancy clusters. The S parameter remains significantly enhanced at the periphery even after 25 revolutions. Hence, contrary to dislocation density spatial distribution of vacancy clusters is far from being uniform even after prolonged high pressure torsion deformation.

Journal of Physics: Conference Series **674** (2016) 012001 doi:10.1088/1742-6596/674/1/012001

1. Introduction

Severe plastic deformation (SPD) [1] enables to achieve extreme grain refinement down to nanoscopic dimensions leading to the formation of ultra fine grained (UFG) structure [2]. UFG materials with grain size below ~500 nm are characterized by highly non-equilibrium structure with high density of grain boundaries and high density of defects [1]. This leads to unique physical properties, in particular abnormally high diffusion activity [3] and very attractive mechanical properties characterized by high strength and sufficient ductility [4]. High-pressure torsion (HPT) [1,5] is the most efficient SPD technique for grain refinement. The principle of HPT technique is illustrated in Fig. 1a. The sample is placed between two anvils and compressed by a high pressure p. One of the anvils is slowly rotating. Hence the sample is strained by torsion under a high pressure of several GPa. Bulk UFG materials with grain size typically around 100 nm can be produced by this technique [1,5]. Samples prepared by HPT are typically disk shaped with the thickness up to a few mm and the diameter up to 20 mm. In HPT deformation the strain increases with the radial distance r from the centre of the sample corresponding to the rotation axis, see Fig. 1b. The equivalent von Misses strain ε imposed by HPT deformation is given by [6]

$$\varepsilon = \frac{2\pi r N}{h\sqrt{3}}, \qquad (1)$$

where r is distance from the centre, h is the thickness of sample disk and N is the number of HPT revolutions. Because of non-uniform strain during HPT it is important to examine the homogeneity of UFG structure and the spatial distribution of deformation-induced defects across the sample disk.

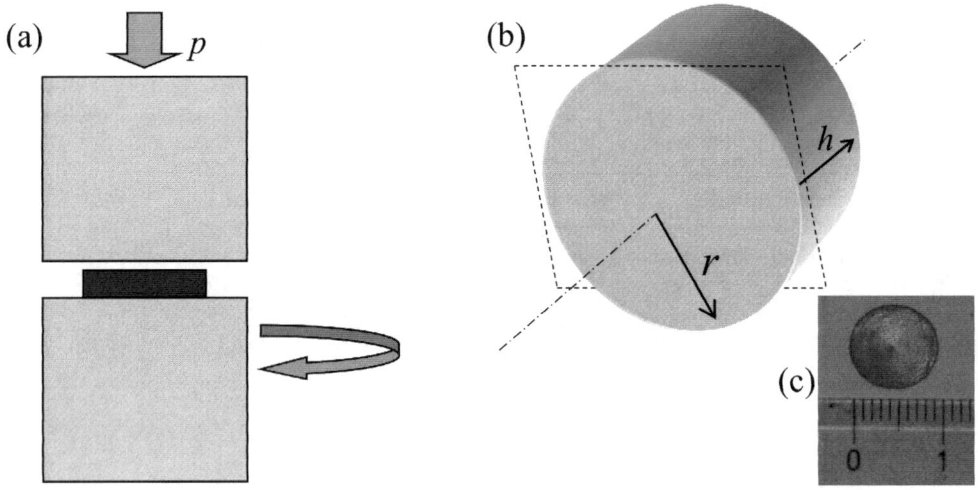

Figure 1. (a) Schematic illustration of HPT deformation; (b) depiction of HPT deformed specimen with diameter r and thickness h; (c) an image of HPT deformed Cu sample studied in this work.

The aim of the present work is to characterize the homogeneity of microstructure and the spatial distribution of defects in HPT-deformed Cu. Positron annihilation spectroscopy (PAS) was employed for mapping of the spatial distribution of defects. This was performed using two PAS techniques: (i) positron lifetime (LT) spectroscopy and (ii) Doppler broadening (DB) of annihilation radiation. The LT spectroscopy enables to identify defects present in the sample and to determine their concentrations. But LT measurements with conventional positron source are time consuming and take usually at least one day. On the other hand measuring period of ~ 30 min is enough to accumulate a DB spectrum with statistics sufficient for precise determination of the line shape parameters S and W. As was demonstrated in Ref. [7] the S parameter can be used for non-destructive mapping of the spatial distribution of defects. In this work spatially resolved PAS was combined with mapping by microhardness testing and direct observation of microstructure by transmission electron microscopy (TEM).

2. Experimental

Cu (99.95% purity) specimens were subjected to room temperature quasi-constrained HPT straining using the compressive pressure of 4 GPa. The number of HPT revolutions was gradually increased from 1 to 25. HPT-deformed specimens were disk shaped with the diameter of ≈ 9 mm and the thickness of ≈ 0.3 mm, see Fig. 1c.

The homogeneity of microstructure was characterized by Vickers microhardness (HV) testing performed using STRUERS Duramin 300 hardness tester with load of 100 g applied for 10 s. HV measurements were performed on a perpendicular grid with incremental spacing of 0.5 mm. Colour-coded HV maps were constructed to provide pictorial displays of microstructure homogeneity across the sample disk. HPT-deformed Cu samples were polished to mirror like quality for HV investigations.

PAS investigations were performed using a $^{22}NaCO_3$ positron source with activity of 1 MBq. The source spot was carefully deposited on a 2 μm thick mylar foil so that the spot diameter was smaller than 1 mm.

A digital spectrometer [8] with time resolution of 145 ps (FWHM of resolution function) was employed for LT spectroscopy. At least 10^7 positron annihilation events were collected in each LT spectrum which was decomposed into exponential components by a maximum likelihood code [9]. A well annealed (850 °C) pure Cu reference sample characterized by a positron lifetime of 114 ps was used to determine the source contribution to the LT spectra. It consisted of two components with lifetimes 368 ps and 1.5 ns and corresponding relative intensities 8 and 1%. These contributions come from positrons annihilated in the $^{22}NaCO_3$ source spot and the covering mylar foils, respectively, and were always subtracted from the spectra. Since it took one day to accumulate single LT spectrum with sufficient statistic it was not feasible to perform two-dimensional mapping of the samples by LT measurements. Instead we made 'line scans' when the positron source was moved along a line from the centre towards the edge and LT measurements were performed at various distances from r from the centre as illustrated in Fig. 2a.

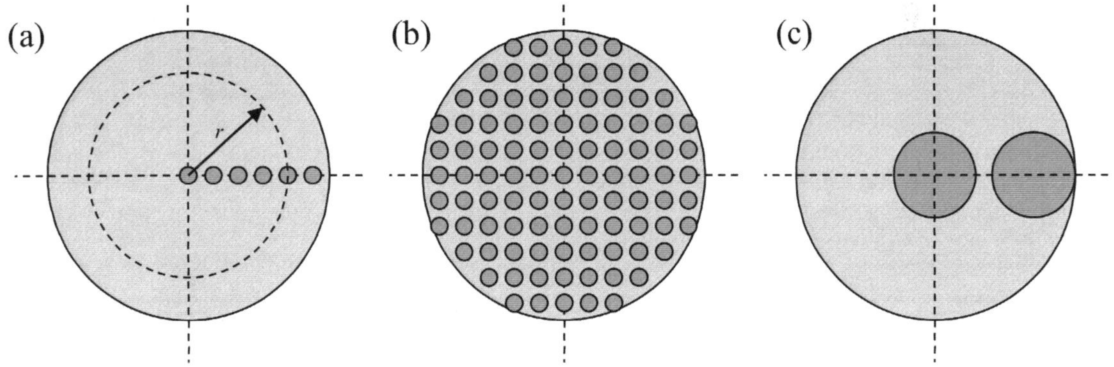

Figure 2. Schematic illustration of (a) line scan performed by LT spectroscopy; (b) DB/HV mapping performed across the whole sample; (c) regions cut from the sample for preparation of the foils for TEM, i.e. the centre region at $r = 0$ and the periphery region centred at $r = 3$ mm. Note that for better visibility Figs. 2(a) and 2(b) are not in scale, the actual mesh was denser.

DB of annihilation radiation was measured by a high purity germanium (HPGe) detector with energy resolution of 1.30 keV at 511 keV. At least 5×10^5 counts were accumulated in the annihilation peak in each spectrum. DB of the annihilation peak was evaluated using the line shape S parameter.

The central region for calculation of the S parameter was chosen symmetrically around the 511 keV annihilation peak from 510.07 to 511.93 keV. All S values presented in this work were normalized to the S parameter S_0 = 0.5045(5) measured in the centre of the Cu sample subjected to one HPT revolution. As shown schematically in Fig. 2b the DB measurements were performed on a perpendicular grid with incremental spacing of 0.5 mm by placing the positron source spot to various nodes of the grid using a micrometer x-y shift. The uncertainty in the position of centre of the positron source was approximately ~ 0.1 mm. It took around 30 min to measure each point. Colour coded maps of S parameter values were constructed in the similar way as in the case of HV mapping. PAS investigations were performed directly on HPT-deformed samples without any surface treatment. In a single HPT-deformed sample selected from the series a sub-surface layer with thickness of ≈30 μm was removed by chemical etching in order to check whether the density of deformation-induced defects changes with depth. The PAS results for the etched sample were virtually the same as those for the sample prior to etching. This testifies that there is no significant change of defect density with the depth from the surface.

TEM observations were carried out on a JEOL JEM 2000 FX electron microscope operating at 200 kV. Foils for TEM were prepared by electrolytic polishing in STRUERS TENUPOL 5 jet-polishing unit using 50% H_3PO_4 at 10 °C. As shown schematically in Fig. 2c foils for TEM investigations were cut (i) from the centre of the sample disk (r = 0) and (ii) from region at the periphery (the centre of the foil was at r = 3 mm) in order to compare the microstructure in these two regions.

3. *Ab-initio* theoretical calculations

Density functional theory was employed for *ab-initio* calculations of lifetimes of positrons trapped at various point defects in Cu. Positron density was calculated within the so-called standard scheme employing the atomic superposition (ATSUP) method [10]. The electron–positron correlation was treated using the local density approximation (LDA) according to the parameterization by Boroński and Nieminen [11]. The calculations were performed in 256 Cu atom based supercells. Point defects (vacancies and vacancy clusters) were modelled simply by removing the corresponding number of atoms from the supercell.

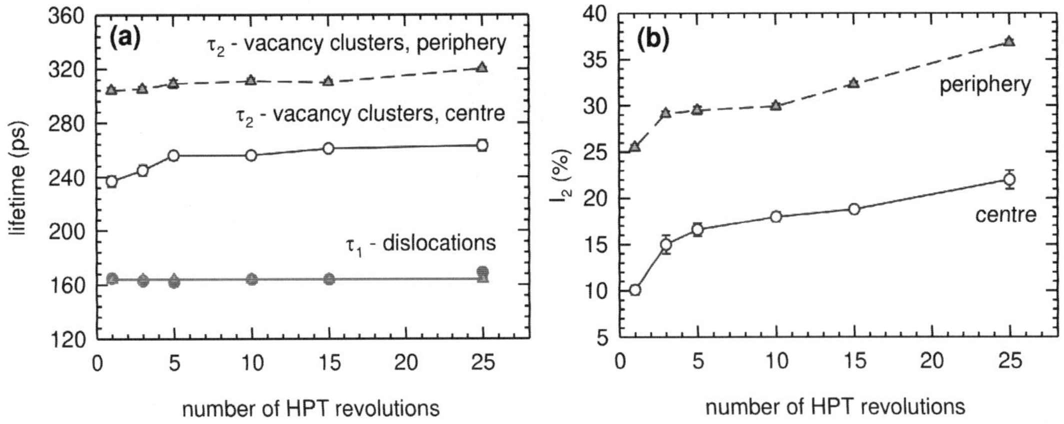

Figure 3. (a) Development of positron lifetimes with increasing number of HPT revolutions; (b) dependence of the intensity I_2 of positrons trapped at vacancy clusters on the number of HPT revolutions. The sample was measured in the centre (r = 0) and at the periphery (r = 3 mm).

4. Results and Discussion

HPT deformed Cu samples exhibit two component LT spectra. Lifetimes of both these components are significantly higher than the bulk Cu lifetime τ_B = 114 ps [12] indicating that both these components come from positrons trapped at defects. No free positron component could be resolved in the LT spectra. This testifies to the very high density of defects even after the first HPT revolution leading to saturated positron trapping.

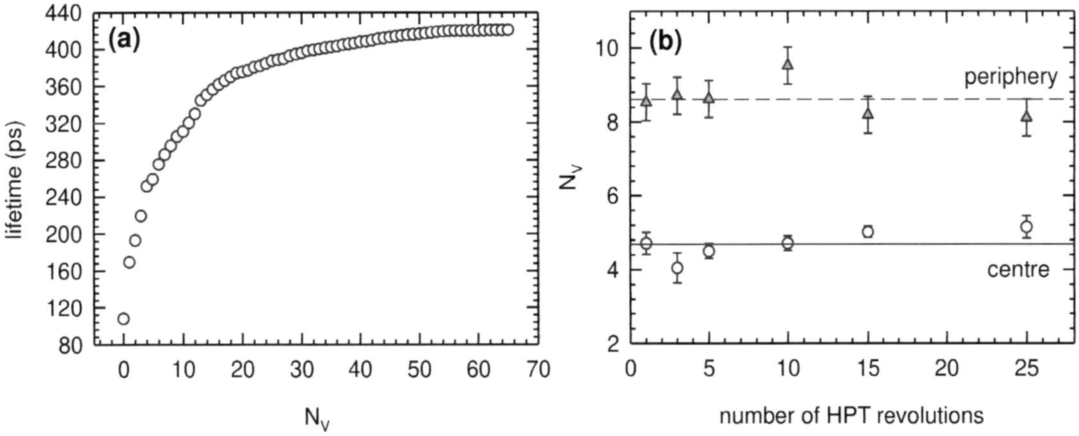

Figure 4. (a) Calculated positron lifetime for vacancy clusters of various sizes plotted as a function of the cluster size (expressed as the number of vacancies N_V in the cluster); (b) dependence of the mean size of vacancy clusters on the number of HPT revolutions. The sample was measured in the centre ($r = 0$) and at the periphery ($r = 3$ mm).

The shorter component with lifetime $\tau_1 \approx 165$ ps represents a contribution of positrons trapped at dislocations [12,13]. Thus, dislocations created during SPD represent the dominating type of defects in HPT-deformed Cu samples. The longer component with lifetime τ_2 falling in the range 240 – 320 ps can be attributed to positrons trapped at vacancy clusters created by agglomeration of vacancies [13,14]. Plastic deformation introduces vacancies by non-conservative motion of dislocations. The most common mechanism of vacancy creation during plastic deformation is climbing of edge dislocations which results in the formation of chains of vacancies. Since vacancies in Cu are mobile at room temperature [15] they either disappear by diffusion to sinks at grain boundaries and surface or agglomerate into small vacancy clusters which are stable at room temperature.

The development of dislocations and vacancy clusters during HPT processing is presented in Fig 3. Lifetimes τ_1 and τ_2 resolved in LT spectra are plotted in Fig. 3a as a function of the number N of HPT revolutions. The lifetime of the dislocation component remains constant during HPT processing except of statistical scattering while the lifetime τ_2 of the cluster contribution slightly increases with increasing number of HPT revolutions. The relative intensity I_2 of the cluster contribution plotted in Fig. 3b increases with increasing number of HPT rotations as well. The samples were measured in the centre ($r = 0$) and also at the periphery ($r = 3$ mm). Both the lifetime τ_2 and the intensity I_2 of the cluster component are remarkably higher at the periphery than in the centre. This testifies that vacancy clusters at the periphery are larger than in the centre.

In order to determine the size of vacancy clusters experimental data were compared with *ab-inito* theoretical calculations. Fig. 4a shows the calculated lifetime of positrons trapped at vacancy clusters of various sizes. The size of vacancy clusters is expressed as number of vacancies contained in the cluster. The lifetime of trapped positrons obviously increases with increasing cluster size. For small vacancy clusters positron lifetime steeply increases while for large clusters in gradually saturates. By comparing the experimental lifetime τ_2 with the theoretical calculations in Fig. 4a the mean size of vacancy clusters in HPT Cu samples was determined.

Fig. 4b shows the dependence of the mean size of vacancy clusters on the number of HPT turns. Vacancy clusters at the centre consist of 4–5 vacancies, i.e. they are smaller than the clusters at the periphery, which consist of 8–9 vacancies. No significant changes in the size of vacancy clusters with increasing number of HPT turns were observed.

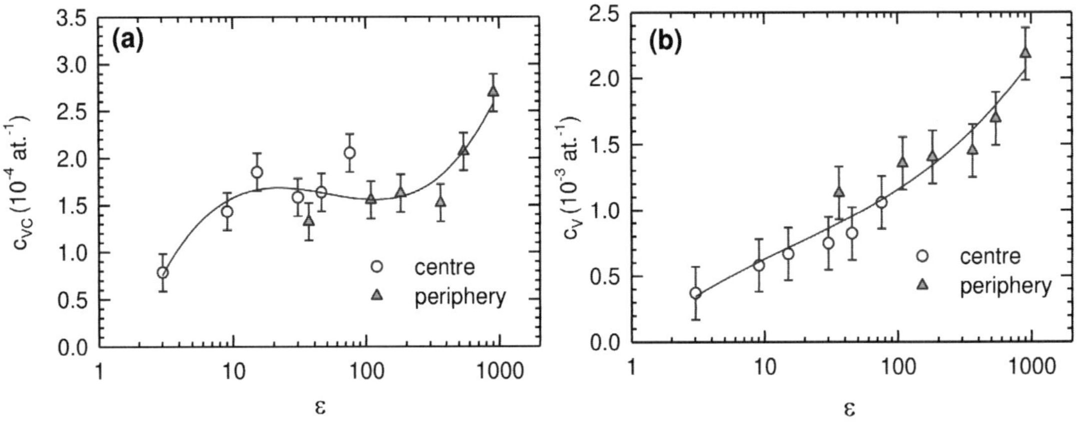

Figure 5. The concentration of vacancy clusters (a) and the concentration of deformation-induced vacancies which agglomerated into clusters (b) plotted as a function of the equivalent strain ε. Solid lines are intended to guide the eye only.

The vacancy production rate Π at low temperature plastic deformation is governed by the strain rate $\dot{\varepsilon}$ [16]

$$\Pi = \alpha \frac{\sigma \Omega_0}{E_f} \dot{\varepsilon}, \qquad (2)$$

where the coefficient $\alpha \approx 0.1$ [17], σ is the applied stress, Ω_0 the atomic volume and E_f the vacancy formation energy. Since in torsion deformation the strain rate increases with the radial distance from the centre of the sample the production rate of vacancies at the periphery is higher than in the centre. This leads to the formation of bigger vacancy clusters at the periphery.

Because of saturated positron trapping the ratio I_2/I_1 of the relative intensities of positrons trapped in vacancy clusters and at dislocations is directly proportional to the ratio of defect concentrations:

$$\frac{I_2}{I_1} = \frac{\nu_{VC} c_{VC}}{\nu_D \rho_D}, \qquad (3)$$

where c_{VC} denotes the concentration of vacancy clusters and ρ_D stands for the dislocation density. The symbols ν_{VC} and ν_D denote the specific positron trapping rate to vacancy clusters and dislocations, respectively. The specific positron trapping rate for dislocations in Cu $\nu_D = 0.6 \times 10^{-4}$ m^2 s^{-1} was estimated in Ref. [18]. The specific positron trapping rate for a small vacancy cluster consisting of N_V vacancies ($N_V \leq 10$) is proportional to the number of vacancies in the cluster. Hence, Eq. (3) can be rewritten as

$$\frac{I_2}{I_1} = \frac{N_V \nu_V c_{VC}}{\nu_D \rho_D}, \qquad (4)$$

where $\nu_V = (1.2 \pm 0.2)\ 10^{14}$ at. s^{-1} [19] is the specific positron trapping rate for a monovacancy in Cu. The dislocation density ρ_D in HPT-deformed Cu samples was determined in Ref. [13] by X-ray line profile analysis and was found to increase with increasing number of HPT revolutions. Using the ρ_D values determined in Ref. [13] the concentration of vacancy clusters was calculated from Eq. (4) and is plotted as a function of the equivalent strain ε in Fig. 5a. One can see in the figure that c_{VC} increases at low strains ($\varepsilon \leq 10$), then it exhibits a plateau followed by further increase at very high strains ($\varepsilon \geq 200$). Increasing number of deformation-induced vacancies leads either to increase of the average size of vacancy clusters and/or to increase of their concentration. However, only the latter effect is visible in Fig. 5a. Hence, in the 'plateau region' ($10 < \varepsilon < 200$) the increasing concentration of deformation-induced vacancies most probably leads to the formation of bigger vacancy clusters without significant change of their concentration. Since the vacancy clusters consist on average of N_V vacancies the net concentration of deformation-induced vacancies which agglomerated into clusters was estimated as $c_V \approx N_V\, c_{VC}$ and is plotted in Fig. 5b as a function of the equivalent strain ε. One can see in the figure that c_V continuously increases with increasing strain testifying that plastic deformation applied during HPT processing introduces continuously vacancies into the sample.

Hence, there are two factors influencing vacancy clusters in HPT-deformed Cu: (i) the strain rate and (ii) the total equivalent strain. Higher strain rate leads to formation of bigger vacancy clusters due to higher production rate of vacancies. Higher equivalent strain leads to higher concentration of deformation-induced vacancies and consequently to formation of vacancy clusters which are larger in size or are present in higher concentration.

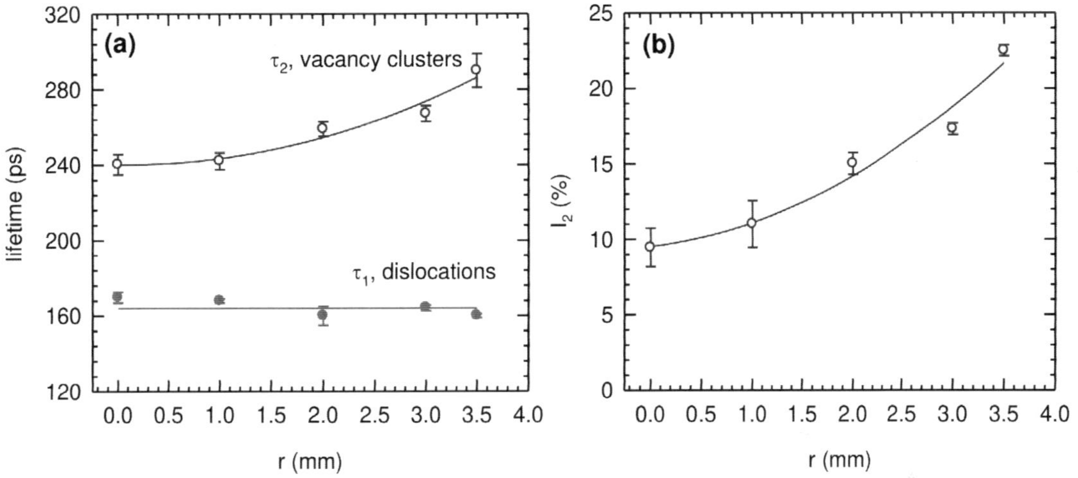

Figure 6. Results of a 'line scan' performed by LT spectroscopy across the radius of UFG Cu deformed by to 2 HPT revolutions: dependence of (a) lifetimes resolved in LT spectra and (b) the relative intensity I_2 of positrons trapped at vacancy clusters on the radial distance r from the centre of the sample disk. Solid lines are intended to guide the eye only.

Detailed mapping of spatial distribution of defects across the radius of sample was performed by LT spectroscopy for the UFG Cu sample deformed by 2 HPT turns. Fig. 6a shows lifetimes resolved in LT spectra plotted as a function of the radial distance from the centre. Saturated trapping at dislocations and in vacancy clusters was observed in all points measured. The dependence of the relative intensity I_2 of positrons trapped at vacancy clusters on the radial distance from the centre is shown in Fig. 6b. Obviously both the lifetime τ_2 and the intensity I_2 of the cluster contribution gradually increase with increasing distance from the centre. Hence, the average size of vacancy clusters gradually increases from the centre of the sample towards the edge due to increasing production rate of vacancies. Since larger vacancy clusters are more efficient positron traps characterized by higher specific positron trapping rate, the fraction of positrons trapped at these clusters increases with r even if the concentration of vacancy clusters remains unchanged.

Fig. 7 shows results of HV mapping of UFG Cu samples subjected to various numbers N of HPT revolutions. The HV maps exhibit axial symmetry with respect to the centre of the sample as expected for deformation by torsion. One can see in the figure that HV at the periphery quickly increases and becomes saturated at the value HV ~ 155. On the other hand, HV in central region of the sample is significantly lower. With increasing number of HPT revolutions the region with enhanced hardness gradually extends from the periphery towards the centre of the sample disk. Finally in the sample deformed by 25 HPT revolutions HV becomes almost uniform practically across the whole sample except for a small region in the centre of the sample.

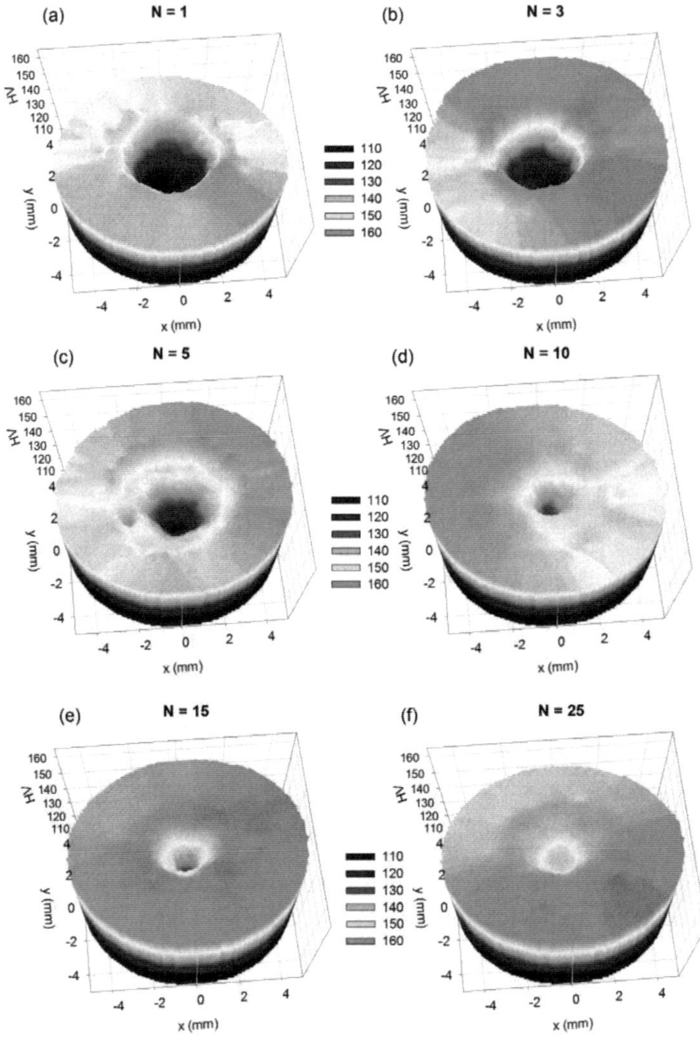

Figure 7. Colour coded map constructed from HV values measured on a rectangular grid across the sample disk for UFG Cu deformed by various numbers N of HPT revolutions.

The average HV at various distances from the centre of the sample were calculated by angular averaging the values measured at the same distances from the centre. Fig. 8a shows the dependence of HV on the equivalent strain ε. The HV values for samples subjected to various numbers of HPT turns fall on a common sigmoid 'master curve' when plotted versus the equivalent strain. Small scattering between various samples is likely caused by uncertainties of centring the sample disk during HPT processing. One can see in the figure that HV increases with increasing strain and gradually saturates at HV ~ 155 for $\varepsilon \geq 200$.

Microstructure of samples deformed by various numbers of HPT revolutions was examined by TEM in order to correlate microstructure evolution with hardness variations. Each sample was probed

Journal of Physics: Conference Series **674** (2016) 012001 doi:10.1088/1742-6596/674/1/012001

in the centre and at the periphery as shown in Fig. 2c. TEM micrographs for the sample deformed by single HPT revolution are presented in Fig. 9. Heavily deformed structure with a high density of dislocations and a strong stress was observed both in the centre and at the periphery. Microstructure in the centre exhibits features typical for initial stages of dislocation re-arrangement, i.e. the formation of cells with low dislocation density separated by dislocation walls containing significantly higher dislocation density. On the other hand, more refined microstructure was observed at the specimen periphery. Dislocation walls become sharper and transform into sub-grain boundaries. Several grains with dislocation-free grain interiors separated by sharp grain boundaries can be observed in Fig. 9b. Hence, TEM investigations revealed significantly different microstructure in the centre and at the periphery. Stronger grain refinement at the periphery is in agreement with higher strain and explains also enhanced hardness at the periphery.

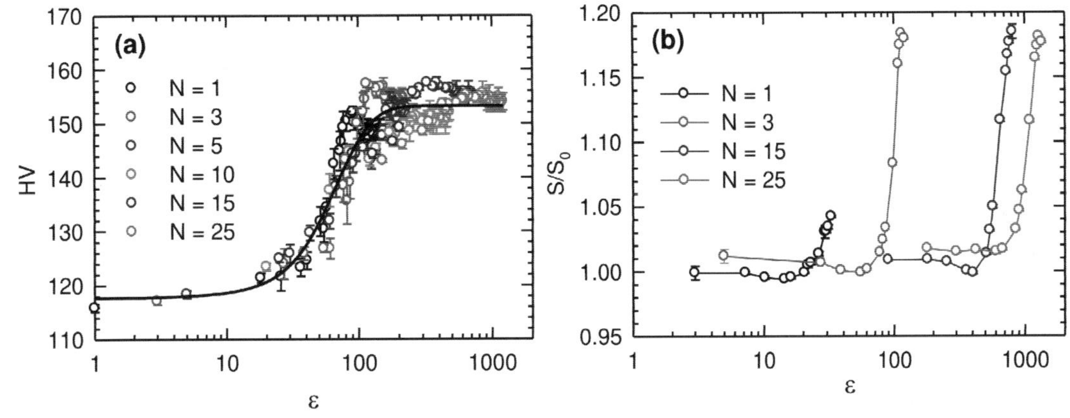

Figure 8. The dependence of HV (a) and the *S* parameter (b) on the equivalent strain ε. Solid lines are guides for the eye.

The microstructure of the sample subjected to 3 HPT revolutions is shown in Fig. 10. The centre of the sample (Fig. 10a) is characterized by dislocation cells and partially developed sub-grains, i.e. the microstructure similar to that observed in the centre after one HPT turn, cf. Fig. 9a. Two types of grains can be recognized at the periphery of the sample (Fig. 10b): (i) almost dislocation-free grains with sharp equilibrium grain boundaries and (ii) grains separated by non-equilibrium grain boundaries with diffuse contrast. The mean grain size at the periphery was estimated as ~ 500 nm.

Fig. 11a shows the microstructure in the centre of sample subjected to 15 HPT turns. The microstructure in the centre consists of dislocation cells and sub-grains separated by tangled dislocations. The periphery of the sample shown in Fig. 11b exhibits refined structure with the mean grain size of 200-300 nm. Similarly to the sample subjected to 3 HPT revolutions one can distinguish in this specimen grains without dislocations with sharp grain boundaries and grains containing dislocations separated by non-equilibrium grain boundaries.

Hence, from TEM investigation it becomes clear that microstructure in the centre of sample differs significantly from that at the periphery even after 15 HPT revolutions. X-ray line profile analysis revealed that dislocation density exhibits only moderate variations across the HPT samples [13,20]. Hence, the enhanced HV at the periphery is most probably caused by finer microstructure with higher density of grain boundaries (so called grain boundary strengthening). It has to be noted that hardness is related to the mean grain size *d* by the Hall-Petch relation [21,22] HV ~ $d^{-1/2}$.

Colour coded maps constructed from the *S* parameters measured by DB spectroscopy on the nodes of a rectangular mesh across the sample disks deformed by various number of HPT turns are displayed in Fig. 12. Similarly to HV mapping also the *S* parameter maps exhibit axial symmetry with respect to the centre of the sample. The *S* parameter map for the sample deformed by a single HPT turn is plotted in Fig. 12a. The *S* parameter values are clearly enhanced at the periphery of the sample disk. In the sample deformed by 3 HPT revolutions the *S* parameter at the periphery increases even more and saturates at $S/S_0 \sim 1.18$, see Fig. 12b. Further increase in the number of HPT revolutions leads to increase of *S* also in the central region, but a pronounced difference between the centre and the periphery remains even in the sample deformed by 25 HPT turns, see Fig. 12d.

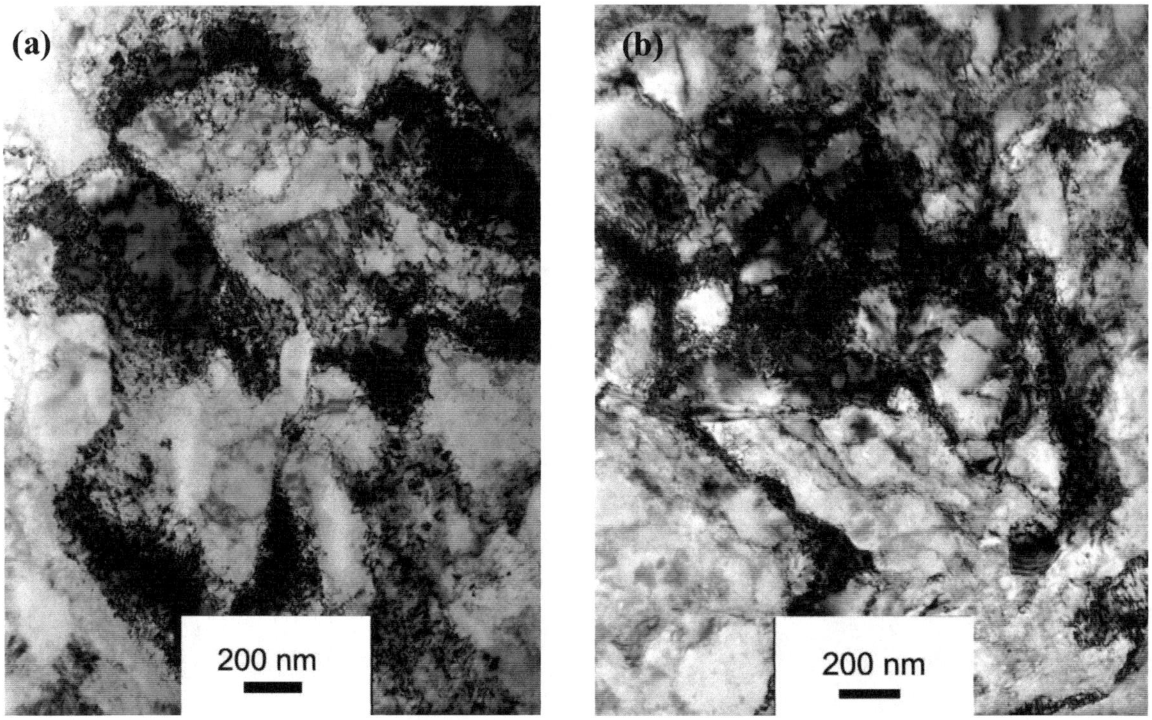

Figure 9. Bright field TEM micrographs of Cu subjected by single HPT revolution: (a) centre of the sample ($r = 0$); (b) periphery ($r = 3$ mm).

The HV and *S* parameter mapping should be considered as mutually complementary techniques providing different point of views on the UFG microstructure. HV is influenced mainly by the grain size (grain boundary strengthening) and the dislocation density (work hardening) while DB spectroscopy is sensitive to the spatial distribution of dislocations and also to vacancy clusters which have relatively low impact on hardness. The general trend in HV and *S* parameter is similar – i.e. both increase with increasing radial distance from the centre due to enhanced defect density and decreasing grain size. But different sensitivity of HV and DB technique leads to remarkable differences between HV and *S* parameter maps which can be clearly seen by comparing Figs. 7 and 12. The sample subjected to 25 HPT revolutions is characterized by almost uniform HV, see Fig. 7f, while the *S* parameter in this sample remains significantly higher at the periphery compared to the centre, see Fig. 12d. This difference is mainly due to vacancy clusters which are substantially bigger at the periphery than in the centre due to higher production rate of vacancies. Hence, although HV mapping showed only a slight difference between the centre and the periphery in the sample subjected to 25 HPT revolutions, DB mapping revealed that lateral distribution of vacancy clusters is far from being uniform and clusters formed at the periphery are always bigger than in the centre.

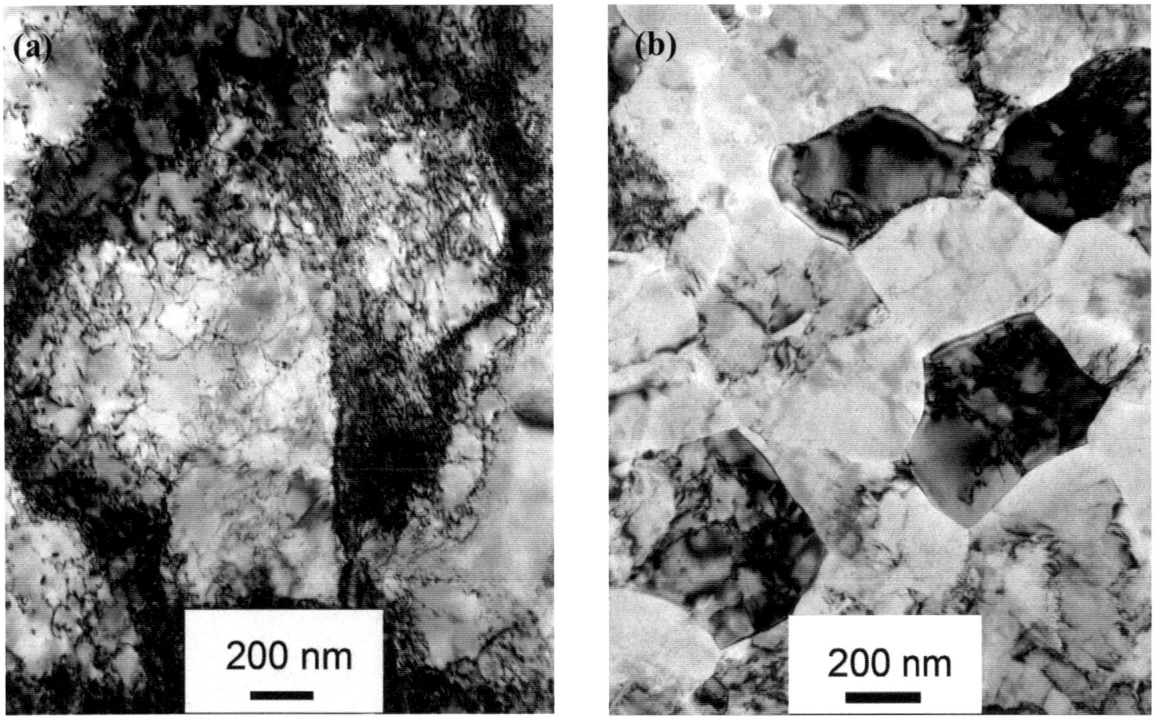

Figure 10. Bright field TEM micrographs of Cu deformed by $N = 3$ HPT revolutions: (a) centre of the sample ($r = 0$); (b) periphery ($r = 3$ mm).

The S parameters obtained for each sample by angular averaging of the values measured at the same distance from the centre are plotted as a function of the equivalent strain in Fig. 8b. By comparing of Figs. 8a and 8b it is clear that there is a striking difference between the development of HV and S parameter values with strain. While HV values for all samples subjected to various number of HPT revolutions fell on a common 'master curve' the S parameter behaves in a completely different way. For each sample S strongly increases at the periphery. This increase is less pronounced in the sample subjected to a single HPT revolution. However, in samples deformed by $N \geq 3$ HPT turns the S parameter behaviour is almost the same independently of strain. Hence, except of a single common 'master curve', the S parameter values when plotted versus the total strain show separate curves for each sample. This indicates that variations of the S parameter across the HPT-deformed samples are predominantly influenced by changes in the size of vacancy clusters. As it was explained in the previous discussion the size of vacancy clusters is determined mainly by the strain rate, which depends on the distance from the centre of the sample irrespective of the total strain. Thus, S parameter maps reflect predominantly variations of the size and the concentration of vacancy clusters while HV maps reflect mainly variations in the grain size

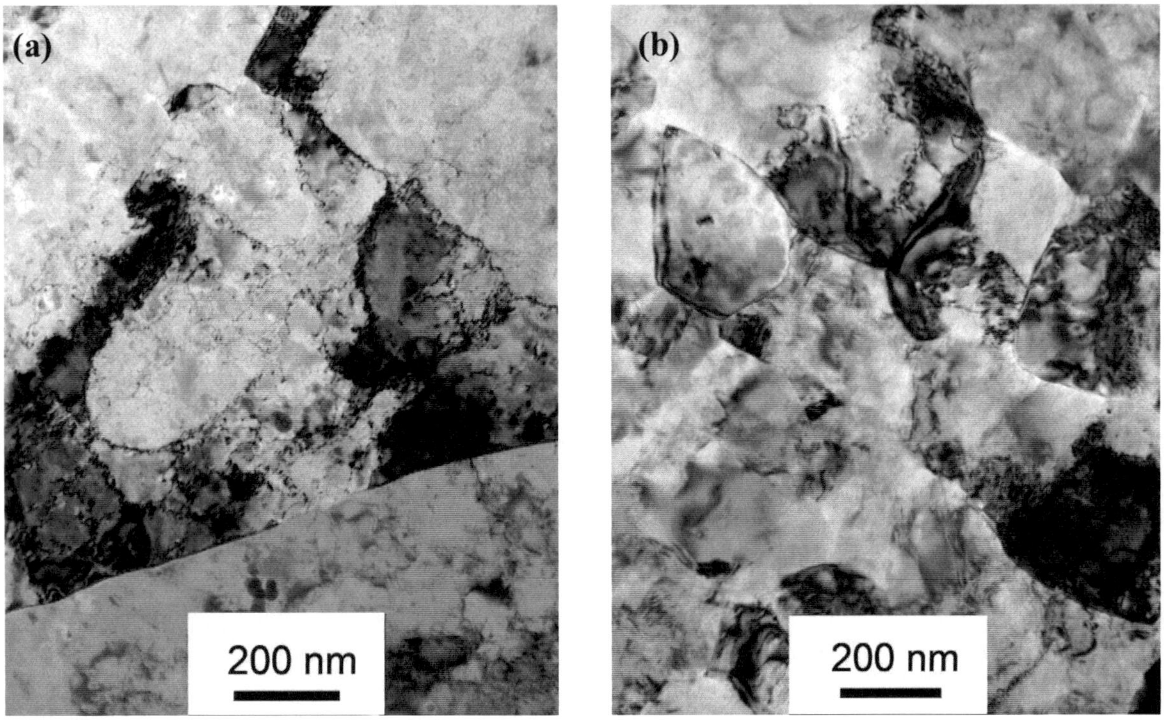

Figure 11. Bright field TEM micrographs of Cu deformed by N = 15 HPT revolutions: (a) centre of the sample (r = 0); (b) periphery (r = 3 mm).

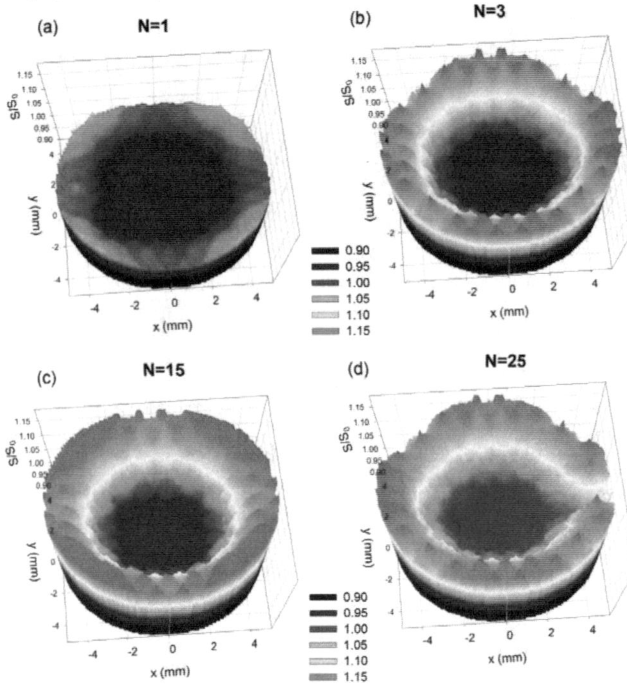

Figure 12. Colour coded maps of S parameters constructed from values measured on a rectangular grid across the sample disk for UFG Cu deformed by various numbers N of HPT revolutions.

5. Conclusions

Two types of lattice defects were found in ultra fine grained Cu prepared by high pressure torsion: (i) dislocations and (ii) small vacancy clusters formed by agglomeration of deformation-induced vacancies. The size of vacancy clusters increases from centre of the sample towards its periphery due to increasing strain rate leading to higher production rate of vacancies. Homogeneity of ultra fine grained structure and spatial distribution of defects were mapped by spatially resolved microhardness testing and positron annihilation spectroscopy. Both these techniques are complementary and reflect different aspects of the microstructure. Hardness mapping visualizes predominantly variations of the grain size while mapping by the S parameters shows mainly changes in the size and the concentration of vacancy clusters. The specimen deformed by 25 revolutions exhibits practically uniform hardness across the whole sample disk while the S parameter is enhanced at the periphery and lowered in the central region. This testifies that the mean grain size became comparable in the whole sample after sufficient number of revolutions but vacancy clusters are always bigger at the periphery than in the centre.

Acknowledgements
This work was supported by the Czech Science Foundation (project P108-13-09436S). One of the authors (M.J.) acknowledges financial support by the same Foundation under the project 14-36566G.References

References

[1] Valiev R Z, Islamgaliev R K and Alexandrov I V 2000 *Prog. Mater. Sci.* **45** 103
[2] Pippan R, Scheriau S, Taylor A, Hafok M, Hohenwarter A and Bachmaier V 2010 *Annu. Rev. Mater. Res.* **40** 319
[3] Valiev R Z, Musalimov R S and Tsenev N K 1989 *Phys. Status Solidi* A **115** 451
[4] Valiev R Z 1998 *Mater. Sci. Eng.* A **234-236** 59
[5] Zhilyaev A P and Langdon T G 2008 *Prog. Mater. Sci.* **53** 893
[6] Valiev RZ, Ivanisenko YV, Rauch EF and Baudelet B 1996 *Acta Mater* **44** 4705
[7] Čížek J, Melikhova O, Janeček M, Srba O, Barnovská Z, Procházka I and Dobatkin S 1998 *Scripta Mater.* **65** 171
[8] Bečvář F, Čížek J, Procházka I and Janotová J 2005 *Nucl. Instrum. Methods* A **539** 372
[9] Procházka I, Novotný I, Bečvář F 1997 *Mater. Sci. Forum* **255-257** 772
[10] Puska M J and Nieminen R M 1983 *J. Phys.* F **13** 333
[11] Boroński E and Nieminen R M 1986 *Phys. Rev.* B **34** 3820
[12] Čížek J, Procházka I, Cieslar M, Kužel R, Kuriplach J, Chmelík F, Stulíková I, Bečvář F and Melikhová O 2002 *Phys. Rev.* B **65** 094106
[13] Čížek J, Janeček M, Srba O, Kužel R, Barnovská Z, Procházka I and Dobatkin S 2011 *Acta Mater.* **59** 2322
[14] Čížek J, Melikhova O, Barnovská Z, Procházka I and Islamgaliev R K 2013 *J. Phys.: Conf. Series* **443** 012008
[15] Mantl S and Trifthäuser W 1978 *Phys. Rev.* B **17** 1645
[16] Militzer M, Sun W P and Jonas J J 1993 *Acta Mater.* **42** 133
[17] Gottstein G, Bewerunge J, Mecking H and Wollenberger H 1975 *Acta Met.* **23** 641
[18] Čížek J, Procházka I, Vostrý P, Chmelík F and Islamgaliev R K 1999 *Acta. Phys. Polonica* A **95** 487
[19] Kluin J E and Hehenkamp Th 1991 *Phys Rev* B **44** 11597
[20] Čížek J, Procházka I, Brauer G, Anwand W, Kužel R, Cieslar M, Islamgaliev R K 2003 *Phys. Stat. Sol.* A **195** 335
[21] Hall E O 1951 *Proc. Phys. Soc. London* **64** 747
[22] Petch N J 1953 *J. Iron Steel Inst. London* **173** 25

Subcascade formation ratio in neutron-irradiated stainless steels

T Yoshiie[1], Y Satoh[2], S S Huang[1], M Horiki[1], K Sato[1] and Q Xu[1]

[1]Research Reactor Institute, Kyoto University, Kumatori, Osaka, Japan
[2]Institute for Materials Research, Tohoku University, Sendai, Japan

E-mail: yoshiie@rri.kyoto-u.ac.jp

Abstract. High-energy-particle irradiation in metals produces cascade damage. If the particle energy is high enough, a cascade is divided into subcascades. In each subcascade, a vacancy rich area is surrounded by an interstitial area. Vacancy clusters are expected to form directly in the vacancy rich area. In this study, the vacancy cluster formation ratio in subcascades was estimated by positron annihilation lifetime spectroscopy and transmission electron microscopy in commercial stainless steels and their model alloys. The vacancy cluster formation ratio was 1.7×10^{-3} and 9.1×10^{-5} in austenitic stainless steel and ferritic/martensitic stainless steel, respectively.

1. Introduction

The defect evolution in materials under high-energy-particle irradiation is a multiscale phenomenon. Defect evolution originates from nuclear reactions (~ 10^{-22} s), followed by point defect formation (~ 10^{-13} s) and clustering (> 10^{-10} s). Point defect clusters and precipitates change the long-term aging properties of materials. The degradation of mechanical properties is a particularly serious problem for the lifetime of nuclear systems. In order to obtain a unified knowledge of the mechanisms of microstructural development during high-energy-particle irradiation, it is important to understand the collision processes that lead to the formation of localized defects.

High-energy-particle irradiation in metals causes cascade damage, where highly concentrated vacancies are surrounded by interstitials. A high-energy cascade produces subcascades, as shown in Figure 1. In each subcascade, there also exists a vacancy rich area surrounded by an interstitial rich area. In the vacancy rich area, the clustering of vacancies occurs. For most FCC metals, stacking fault tetrahedra (SFT) are mainly formed as vacancy clusters near room temperature. Groups of SFT have been observed by transmission electron microscopy (TEM) [1, 2]. It has been shown that these SFT are formed directly from subcascades, from which the cascade size and the number of subcascades have been estimated [1]. However, the cascade structure of austenitic stainless steels with the FCC structure has not been reported. Furthermore, for BCC metals it is difficult to detect subcascades owing to the lack of SFT formation. Instead of SFT, subcascades in BCC metals lead to micro-voids. However, micro-voids less than 1.5 nm cannot be observed by TEM. Therefore, the understanding of subcascades is lacking in these cases and most information is obtained solely from computer simulations, mainly using molecular dynamics (MD) [3].

Austenitic and ferritic stainless steels are both important nuclear materials. The former have been used primarily as nuclear reactor core materials owing to their high corrosion resistance. Ferritic steels are expected to be used as structural materials in future fusion and fast breeder reactors.

The positron is the only probe that can sensitively detect vacancy-type defects in most materials [4, 5]. If there are open spaces in the lattice, such as vacancies where positively charged nuclei are absent, positrons become briefly trapped in these spaces, resulting in longer lifetimes. The increased lifetime depends on the type and size of defects [6-8], and therefore conveys important information regarding vacancy-type defects.

In this study, the subcascade formation ratio in neutron-irradiated austenitic and ferritic/martensitic stainless steels was studied using TEM and positron annihilation lifetime measurements.

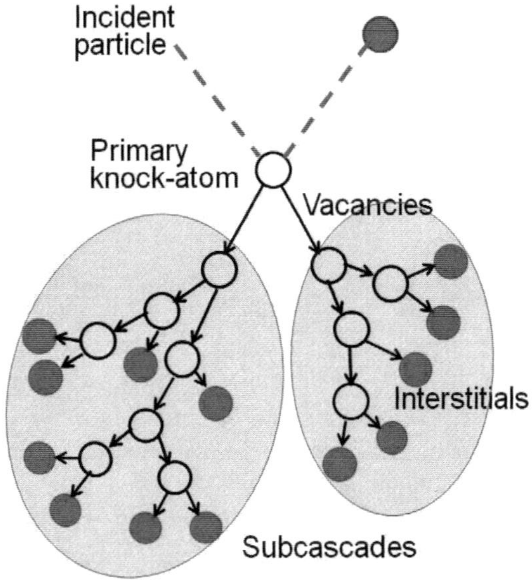

Figure 1. Schematic illustration of subcascades in a cascade.

2. Experimental procedures

The following stainless steel samples were compared in this study: austenitic stainless steel of composition Fe-15.27Cr-15.8Ni-1.88Mn-2.66Mo-0.53Si-0.24Ti-0.055C-0.024P (Ti added modified SUS316SS: JPCA, hereafter referred to as Ti316), model austenitic steel alloys of composition Fe-15Cr-16Ni and Fe-16Cr-17Ni, ferritic/martensitic stainless steel Fe-8%Cr-2%W-0.2%V-0.04%Ta-0.5Mn-0.09Si-0.09C (F82H, referred to as F/MS here) and Fe-8Cr. Each specimen was rolled to 0.1 mm in thickness, from which 3 mm diameter disks were punched out. Ti316, Fe-15Cr-16Ni, and Fe-16Cr-17Ni discs were annealed for 1 h under a vacuum of 1.5×10^{-4} Pa at 1323 K, 1223 K, and 1223 K, respectively. F/MS and Fe-8Cr disks were annealed under high vacuum at 1313 K for 0.5 h and then tempered at 1013 K for 2 hours followed by air cooling. Fission neutron irradiation was performed at the Research Reactor of Kyoto University (6×10^{-4} dpa and 1.5×10^{-3} dpa at 363 K, 1.6×10^{-3} dpa and 5×10^{-3} dpa at 573 K [9]), the Japan Materials Testing Reactor (JMTR) of the Japan Atomic Energy Agency (0.33 dpa at 373 K and 0.11 dpa at 573 K) and Belgian Reactor 2 (BR-2) of the Belgian Nuclear Research Center (0.2 dpa at 363 K and 563 K). Isochronal annealing experiments were performed for 1 h at each temperature after irradiation.

Microstructural observation was performed by a 200 keV TEM, JEOL JEM-2010 at room temperature. Positron annihilation lifetime measurements were carried out at room temperature using a conventional fast-fast spectrometer with a time resolution of 190 ps, corresponding to the full width at half maximum (FWHM). Each spectrum was accumulated at a total of over 1×10^{6} counts for 12 h. The positron lifetime spectra were analyzed using the PALSfit program [10]. To discriminate defect components from bulk components, the lifetime spectrum was decomposed into two or three possible components, a short lifetime τ_1 and two long lifetimes τ_2 and τ_3, based on a trapping model, after subtracting the radioactive source and background components [10]. To evaluate the size of the vacancy clusters (Vn, where n is the number of vacancies), the experimental results were compared with the calculated positron lifetimes of vacancy clusters. The positron annihilation lifetime of a Fe-14Cr-13Ni model alloy of austenitic stainless steel was calculated based on first principles using Monte Carlo density functional theory in the local density approximation [11]. The calculated lifetimes were 106 ps and 183 ps for positrons annihilated in the matrix and in single vacancies, respectively. These values were almost the same as those of Ni [7, 8], and that is why we used the lifetime of Ni for the identification and estimation of the size of defect clusters such as voids and SFT in Ti316. For the lifetimes of micro-voids in F/MS, those in Fe were used [8]. The lifetimes are summarized in Tables 1 and 2.

Table 1. Positron annihilation lifetimes in Ni [7, 8]. V1 and Vn represent single vacancies and micro-voids of n vacancies, respectively. Two values shown for V5 depended on the orientation of clusters. Sm is a SFT composed of m vacancies.

Size of V clusters	V1	V2	V5	V13	V19	V43	V79	S3	S10	S15	S21	S21	S28
Lifetime (ps)	176	195	264/297	341	371	410	427	183	170	159	155	155	130

Table 2. Positron annihilation lifetimes in Fe [8]. V1 and Vn represent single vacancies and micro-voids of n vacancies, respectively. The two values shown for V2 depended on the orientation of clusters.

Size of V clusters	V1	V2	V5	V9	V15	V27	V51	V59	V65	V137
Lifetime (ps)	180	187/202	246	280	368	396	419	426	427	435

3. Results

3.1. Austenitic stainless steel

The defect structure of Ti316 irradiated at 373 K to a dose of 0.33 dpa is shown in the dark field TEM images of Figure 2, where almost no defects are observed. As a comparison, the defect structure of Fe-Cr-Ni is also shown. In Fe-15Cr-16Ni at 373 K, many SFT and interstitial type dislocation loops are observed. Irradiations at 563 K to a dose of 0.2 dpa and at 573 K to a dose of 11 dpa result in dislocation loops, SFTs, and a few precipitates in Ti316. In Fe-16Cr-17Ni at 563 K to a dose of 0.2 dpa and Fe-15Cr-16Ni at 573 K to a dose of 0.11 dpa, voids are observed in addition to SFT and loops, as shown in Figure 3 of the enlarged photo.

The results of positron annihilation lifetime measurements in Fe-16Cr-17Ni are shown in Figure 4. At 363 K, irradiation to 0.2 dpa results in micro-voids, but the detected intensity was very low, and these voids could not be observed by TEM. At 563 and 573 K, micro-voids are also detected and the intensity is high enough to be observed by TEM. In Ti316, the long lifetimes after irradiation at 363 K and 573 K were somewhat shorter than those for single vacancies. The two or three component analysis was impossible after irradiation to 0.2 dpa at 563 K. These lifetimes correspond to those for vacancies on dislocations, SFT, and interfaces between matrix and precipitates [12]. The annealing behavior of Ti316 irradiated at 363 K to a dose of 0.2 dpa is shown in Figure 6. At 523 K, the lifetime decreased and became similar to that of Ti316 without irradiation. Thus, defects formed at 363 K were not stable. The easy annihilation of defects under 523 K suggests that the lifetime at 363 K is due to single vacancies and vacancies on dislocations.

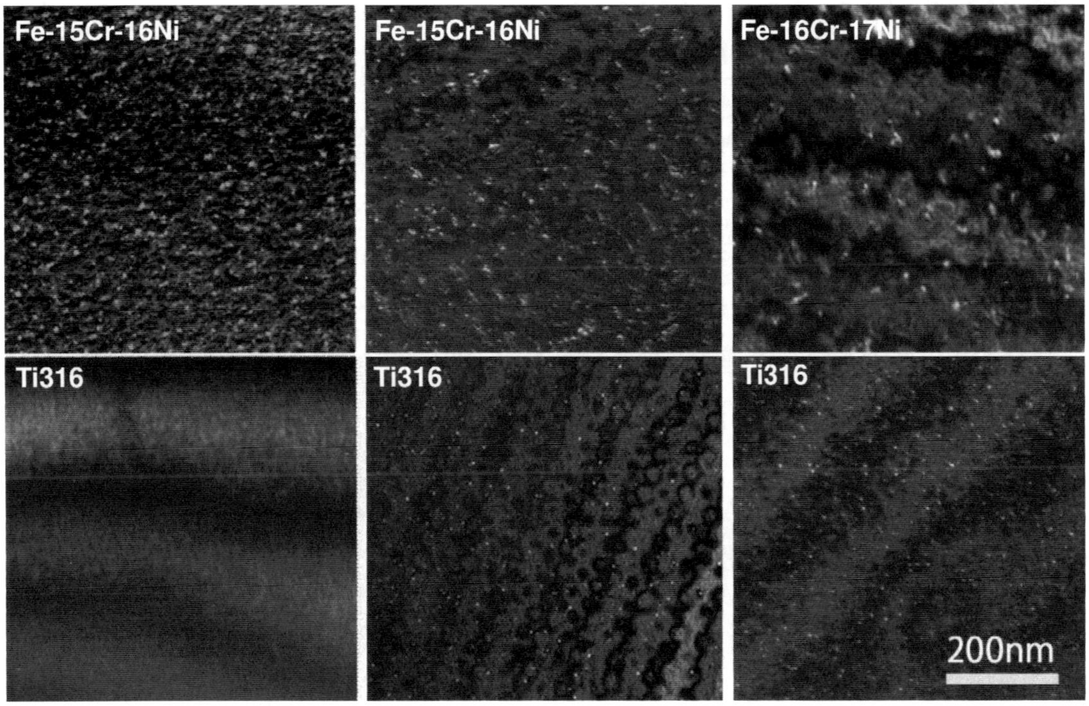

373 K, 0.33 dpa 573 K, 0.11 dpa 563 K, 0.2 dpa

Figure 2. Damaged structures in neutron-irradiated Fe-15Cr-16Ni (373 K and 573 K), Fe-16Cr-17Ni (563 K), and Ti316.

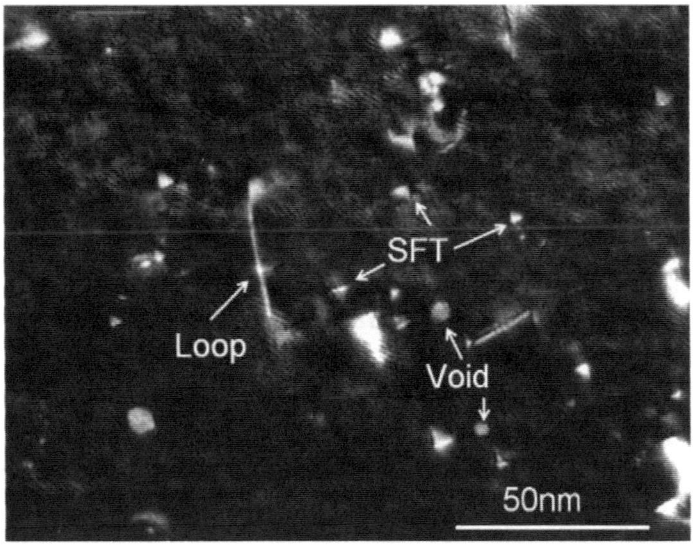

Figure 3. Damaged structures in neutron-irradiated Fe-15Cr-16Ni to a dose of 0.11 dpa at 573 K.

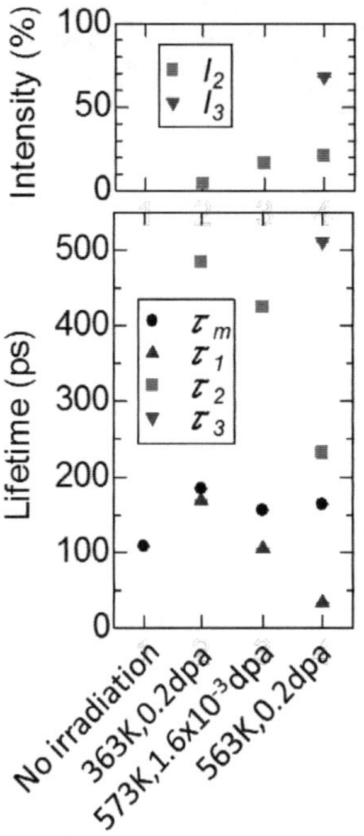

Figure 4. Positron annihilation lifetime measurements of Fe-16Cr-17Ni.

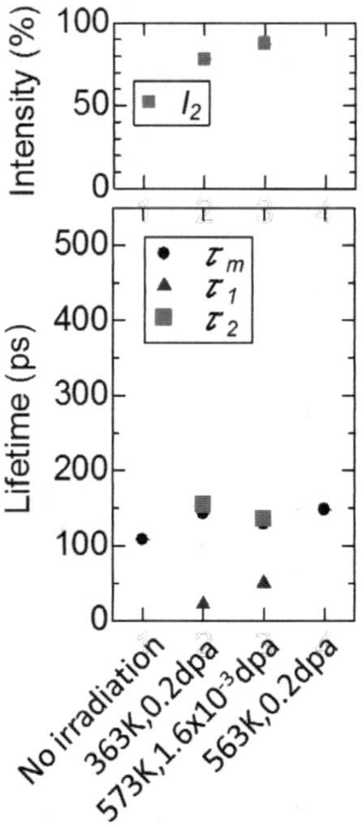

Figure 5. Positron annihilation lifetime measurements of Ti316.

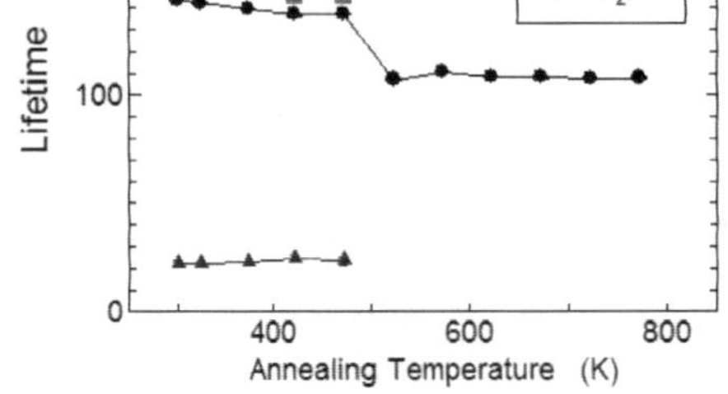

Figure 6. Annealing behavior of neutron-irradiated Ti316 at 363 K to a dose of 0.2 dpa.

3.2. Ferritic stainless steel

Neutron irradiation was conducted for F/MS and Fe-8Cr at 363 K to doses of 6×10^{-4} dpa, 1.5×10^{-3} dpa, and 0.2 dpa, and at 573 K to a dose of 5×10^{-3} dpa. Positron lifetime measurements before irradiation, after neutron irradiation at 363 K and 573 K are shown in Figures 7 and 8. In Fe-8Cr before irradiation, no defects were detected. While in F/MS, a long lifetime of 133 ps was detected. The lifetime can be attributed to dislocations and interfaces between the matrix and precipitates [12]. After irradiation at 363 K in both Fe-8Cr and F/MS, micro-voids were formed. In F/MS, the average size of micro-voids was 11, 14, and 20 vacancies for doses of 6×10^{-4} dpa, 1.5×10^{-3} dpa, and 0.2 dpa, respectively. After

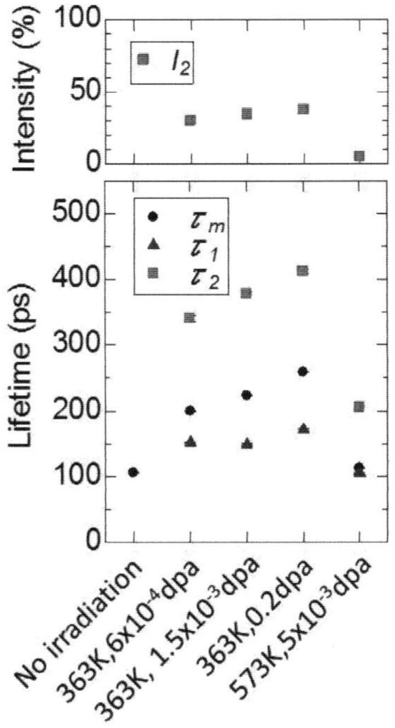

Figure 7. Positron annihilation lifetime measurements of Fe-8Cr.

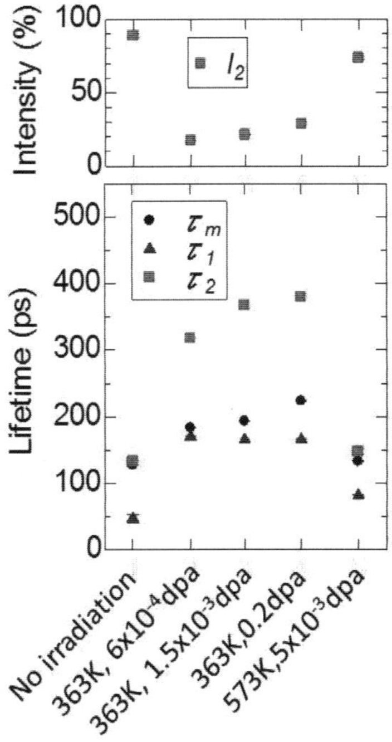

Figure 8. Positron annihilation lifetime measurements of F/MS.

irradiation at 573 K, the mean lifetimes for both alloys were similar to those before irradiation. However, in Fe-8Cr, the lifetime could be decomposed into two components, indicating the formation of defects.

F/MS irradiated at 363 K to 1.5×10^{-3} dpa was isochronally annealed for 1 h at each temperature, as shown in Figure 9. Positron annihilation measurements were conducted after each step at room temperature. A decrease in lifetime was detected after annealing at 573 K. Thus, micro-voids formed at 363 K were not stable at 573 K.

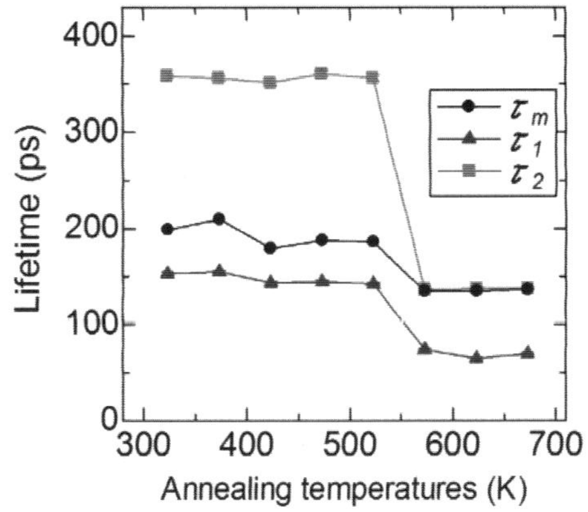

Figure 9. Annealing behavior of neutron-irradiated F/MS at 363 K to 1.5×10^{-3} dpa.

4. Discussion

4.1. Defect structures

The measured τ2 components of positron annihilation lifetime were consistent with TEM observations. In Fe-Cr-Ni irradiated at 363 K and 373 K, vacancies were mobile and accumulated in micro-voids or SFT within vacancy rich areas (subcascades). Vacancies remained in the matrix in TiS16 at 363 K and 373 K, consistent with the fact that vacancies do not move in commercial austenitic stainless steels at 363 K [12]. At 563 K in Fe-16Cr-17Ni and 573 K in Fe-15Cr-16Ni, small SFT annihilated and some SFT and voids grew larger, and could be observed by TEM (Figure 3). In Ti316, only the precipitates grew larger (Figure 5). At this dose range, vacancy-solute diffusion to precipitates prevented vacancy accumulation as micro-voids, as observed in Fe-Cu alloys [13].

In ferritic alloys, no defects were detected in Fe-8Cr prior to irradiation, but in F/MS, precipitates were found. After irradiation at 363 K, micro-voids formed in both Fe-8Cr and F/MS. With increasing dose, the size and concentration of micro-voids increased. For irradiation at 573 K, a few small vacancy clusters, such as V2, remained in Fe-8Cr. These clusters were not stable at 573 K and may form towards the end of the irradiation and remain. For F/MS, void formation was prevented by the growth of precipitates at 573 K, as for Ti316.

The dpa dependence of the trapping rate of the long lifetime at 363 K in F/MS is shown in Table 3, assuming a lifetime of 128 ps (mean lifetime) for F/MS before irradiation. Even before irradiation, defects were present. However, the long lifetime of 133 ps was almost the same as the mean lifetime and we decided that a two state trapping model was appropriate after neutron irradiation at 363 K. The trapping rate is the product of the defect concentration and the capture probability. The capture probability P is calculated from a diffusion model as $P = 4\pi R_T D_+ / \Omega$ [14]. A typical value of 2×10^{-4} m^2/s was used for the diffusion coefficient of positrons, D_+. Ω is the atomic volume. For the trap radius R_T, the radius of micro-voids was used, based on estimates from the long lifetime. With these assumptions, the number density of micro-voids in F/MS at 363 K was 9.8×10^{14}, 1.3×10^{15}, and 1.7×10^{15} /cm^3 for capture probabilities of 9.8×10^{-4}, 1.3×10^{-3}, and 2.1×10^{-3}/s, respectively.

Table 3. Trapping rates of neutron-irradiated F/MS at 363 K.

Irradiation dose (dpa)	6.0×10^{-4}	1.5×10^{-3}	2.0×10^{-1}
Capture probabilities (1/s)	9.8×10^{-4}	1.3×10^{-3}	2.1×10^{-3}
Density of micro-voids (1/cm^3)	9.8×10^{14}	1.3×10^{15}	1.7×10^{15}

4.2. Subcascade analysis

A primary knock-on atom with a large energy makes several well-separated damaged regions, defined as subcascades. Based on the method for estimating subcascade formation energy proposed by Satoh *et al.* [15], the energy for Ni and Fe was calculated. Simulations of atomic displacement cascades are performed using the binary collision approximation for the systematically varied energy of incident atoms to estimate the break up process of a large cascade into subcascades. The mean distance between collisions that transfer a large amount of kinetic energy to a target atom and the size of the vacancy rich region produced by the target atom are compared as a function of the transferred energy to a target atom. The relation between these two parameters is proposed as a criterion for subcascade formation.

The size of the damaged zone, in which a dense population of vacancies is produced (named here the range of vacancies, R) by cascade collisions of a target atom, is calculated using the MARLOWE code [16]. The relationship between the distance of collisions to produce a target atom (D) and R is proposed to serve as a criterion for subcascade formation, as schematically shown in Figure 10. When D is larger than the range R, two damage zones formed from an incident atom do not overlap each other and become separate subcascades (Figure 10(a)). For example, this condition occurs in Fe at an incident energy above 50 keV (Figure 11). D decreases with the decrease of incident energy, and begins to cross R at an incident energy around 10 keV. Below this energy, D is smaller than R, and the damaged zones overlap each other (Figure 10(b)). Here the energy divided into a subcascade is determined to be about 10 keV in Fe. By the same procedure, the subcascade formation energy of Ni is also 10 keV. The energy has not yet been obtained experimentally in Fe.

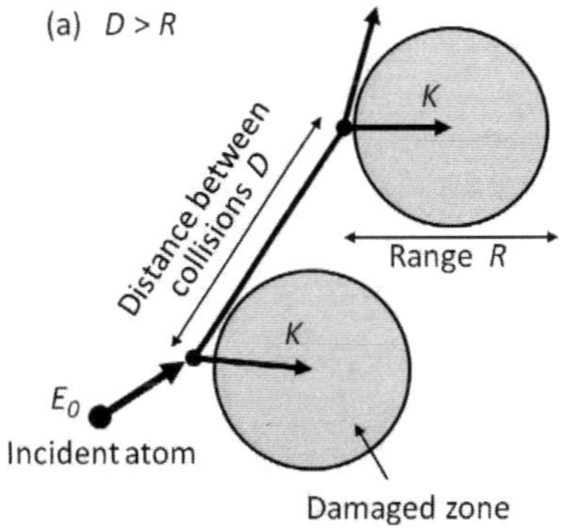

Figure 10. Schematic illustration of the criterion for subcascade formation.

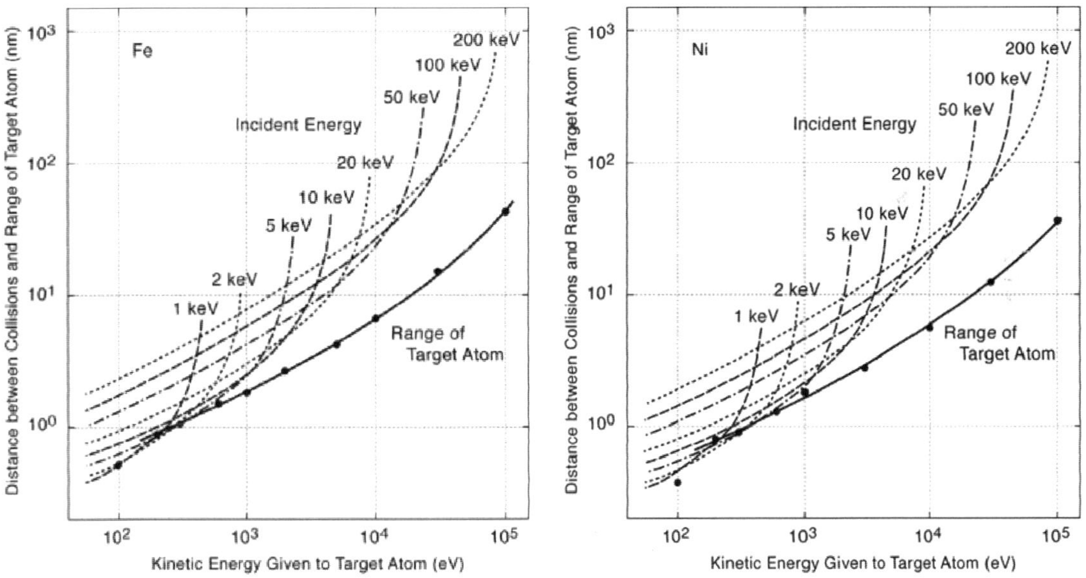

Figure 11. Criterion for the formation of subcascades and subcascade parameters in Fe and Ni. Dotted lines are the mean distance D between collisions in which more energy than K (horizontal axis) is transferred to the target atom. The bolded line shows the range of vacancies R produced by the target atom at its corresponding energy.

The partition of the damage energy K to subcascades is determined by assuming a threshold energy for subcascade formation [1] in an analogous procedure as in the Kinchin-Pease model [17] for Frenkel pair production. A damaged area with a deposited energy above the threshold energy E_{SC} is assumed to evolve into a subcascade. The number of subcascades N_{SC} produced from the primary recoil atom with energy E_{PRA} is

$$\begin{aligned} N_{SC}(E_{PRA}) &= 0 && K(E_{PRA}) < E_{SC} \\ &= 1 && E_{SC} \le K(E_{PRA}) < 2E_{SC} \\ &= \frac{K(E_{PRA})}{2E_{SC}} && 2E_{SC} < K(E_{PRA}), \end{aligned} \quad (1)$$

where $K(E_{PKA})$ is the damage energy for the primary recoil energy E_{PRA}. The cascade formation cross section σ_{CAS} and subcascade formation cross section σ_{SC} are expressed as

$$\sigma_{CAS} = \int_{E_{SC}}^{K_{MAX}} \frac{d\sigma}{dE_{PRA}} dE_{PRA} \quad (2)$$

$$\sigma_{SUB} = \int_{E_{SC}}^{K_{MAX}} N_{SC}(E_{PRA}) \frac{d\sigma}{dE_{PRA}} . \quad (3)$$

The primary recoil energy spectrum $\frac{d\sigma}{dE_{PRA}}$ for the JMTR was calculated using the SPECTER code [18], as shown in Figure 12. In Ni, the total cascade formation cross section was 0.9 barns and the total subcascade formation cross section was 1.6 barns. In Fe, the total cascade and subcascade formation cross sections were 0.86 and 1.3 barns, respectively.

Stoller *et al.* obtained primary damage formation in Fe by MD simulations [3]. In their simulation, 3 large and 3 small damaged areas (subcascades) were formed by 100 keV Fe self-ion irradiation. The number of subcascades in a 100 keV Fe ion in Fe is 5 according to eq (1) with a subcascade formation energy of 10 keV. Their result indicates that a subcascade formation energy of 10 keV in Fe is reasonable.

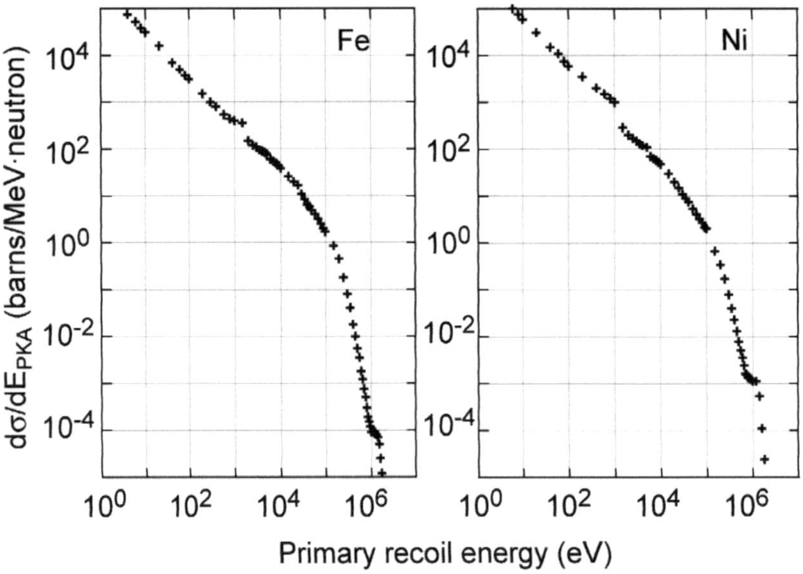

Figure 12. Primary recoil energy spectra of fission neutron-irradiated Fe and Ni as obtained from the SPECTER code [18].

4.3 Sub-cascade formation ratio in stainless steels

In subcascades, vacancy clusters are formed directly, but the production of clusters is not 100%. For Cu, most of the estimated number of subcascades were observed as SFT by TEM [2]. However, for Ni, only a few subcascades were observed as SFT. For austenitic stainless steels, the situation is nearly the same as Ni. Next, we estimate the subcascade formation ratio assuming that SFT were formed in subcascades in Ti316 at 563 K and 573 K, and micro-voids in F/MS at 363 K.

For analyzing the average number density of subcascades (SFT in FCC and micro-voids in BCC), we suppose that there is a volume V around a vacancy cluster in which no additional vacancy clusters can be formed because further cascades eliminate preexisting vacancy clusters through interstitials and simultaneously create new vacancy clusters. In other words, for a number density of preexisting vacancy clusters of ρ, further cascades in a volume ρV per unit volume do not contribute to an increase in the number density of vacancy clusters [19]. The increase of the number density $d\rho$ from an additional fluence $d\varphi$ is expressed as

$$\frac{d\rho}{d\varphi} = N(1 - \rho V)\sigma_{SUB} \tag{4}$$

$$\rho = \frac{1}{V}[1 - \exp(-N\sigma_{SUB}\, V\varphi)], \tag{5}$$

where N is the number density of atoms.

By fitting the dose dependence of SFT in Ti316, V=8.5×10^{-18}cm^3 and σ_{SUB} = 2.7×10^{-3} barns were obtained from eq (5). Figure 13 shows the SFT density and the fit curve using eq (5). The formation ratio of SFT from subcascades was 1.7×10^{-3} by comparing with the subcascade formation cross section of 1.6 barns.

For F/MS, the density of micro-voids obtained in Section 4.1 was used for the analysis. Using the micro-void density ρ in eq (5), V = 5.9×10^{-16} cm^3 and σ = 1.4×10^{-4} barns were obtained. Figure 14 shows the micro-void density and the fit curve using eq (5). The formation ratio of micro-voids from subcascades was 9.1×10^{-5}.

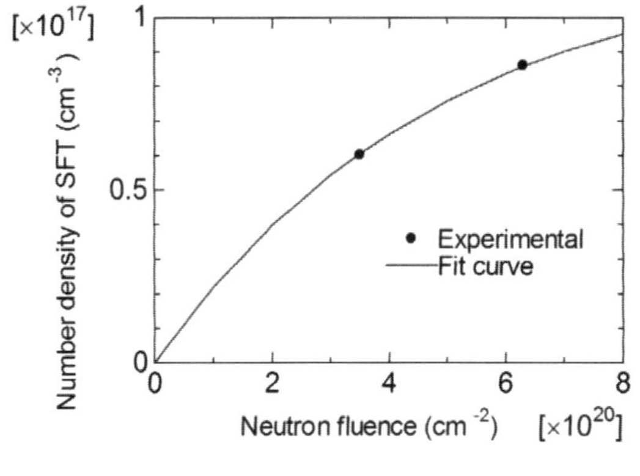

Figure 13. The comparison between experimental results and calculation using eq (5) in Ti316.

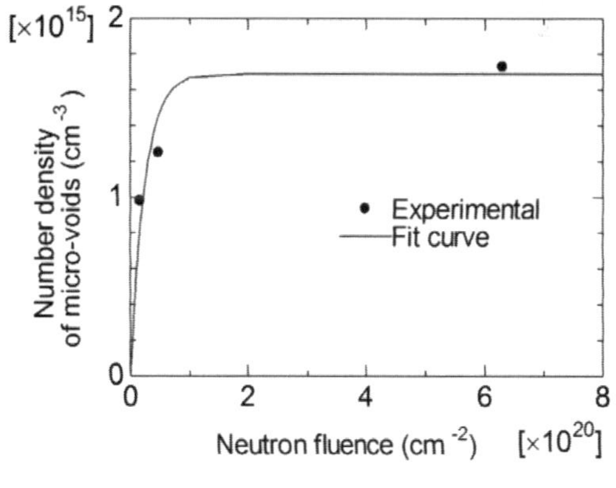

Figure 14. The comparison between experimental results and calculation using eq (5) in F/MS.

Although we obtained values for SFT formation ratio and the micro-void formation ratio, the irradiation facility, irradiation damage rate, and irradiation temperature were different in each irradiation. The damage rate dependence was found to be important for defect evolution [20]. In eq (4), the growth of vacancy clusters directly formed from a subcascade is not taken into account in the present model. The size of the SFT in Ti316 did not change with increasing dose, but in F/MS, micro-voids grew larger at higher doses. These additional conditions should be considered further in future experiments.

Although there were some ambiguities in the experiments and modelling, the subcascade formation ratio was found to be very low in comparison to that in Au and Cu [1,2]. This observation means that clusters formed directly from subcascades are very unstable. This is probably one of the important reasons for the long incubation period for void swelling in these stainless steels [12, 21-23]. Void swelling begins after high dose irradiation, such as several tenths of a dpa.

5. Concluding remarks

The formation of cascades and subcascades are important for determining the initial state of irradiation damage evolution. No subcascade analyses have been performed in commercial steels prior to this study. In BCC metals, this study determines the subcascade formation ratio for the first time. Although computer simulations have undergone remarkable developments, in particular for the study of cascade structures using MD simulations, comparisons between experiments and simulations are important to ensure the reliability of such calculations. The present study directly addresses this important question.

References

[1] Kiritani M, Yoshiie T, Kojima S and Y. Satoh 1990 *Radiation Effect. Defect. Solid.* **113** 75
[2] Satoh Y, Yoshiie T and Kiritani M 1992 *J. Nucl. Mater.* **191-194** 1101
[3] Stoller R E and Greenwood L R 1999 *J. Nucl. Mater.* **271/272** 57
[4] Brandt W and Dupasquier A 1983 *Positron Solid-State Physics* (North-Holland, Amsterdam)
[5] Dupasquier A and Mills A P Jr 1995 *Positron Spectroscopy of Solids* (Amsterdam: IOS Press)
[6] Shivachev B L, Troev T, Yoshiie T 2002 *J. Nucl. Mater.* **306** 105
[7] Kuramoto E, Tsutsumi T, Ueno K, Ohmura M and Kamimura Y 1999 *Comp. Mater. Sci.* **14** 28
[8] Ohkubo H, Tang Z, Nagai Y, Hasagawa M, Tawara Y and Kiritani M 2003*Mater. Sci. Eng.* A **350** 92
[9] Yoshiie T, Hayashi Y, Yanagita S, Xu Q, Satoh Y, Tsujimoto T, Kozuka T, Kamae K, Mishima K, Shiroya S, Kobayashi K, Utsuro M and Fujita Y 2003 *Nucl. Inst. Meth. Phys. Res* A **498** 522
[10] Kirkegaard P, Olsen J V, Eldrup M and Pedersen N J Riso DTU, February 2009, ISBN 978-87-550-3691-8, 44 p, http://palsfit.dk/
[11] Yoshiie T, Cao X Z, Xu Q, Sato K and Troev T D 2009 *Phys. Status Solidi* **C6** 2333
[12] Yoshiie T, Cao X Z, Sato K, Miyawaki K and Xu Q *J.* 2011 *Nucl. Mater.* **417** 968
[13] Yoshiie T, Xu Q and Sato K 2013 *Nucl. Inst. Meth.* B **303** 37
[14] Puska M J and Nieminen R M 1983 *J. Phys.* F **13** 333
[15] Satoh Y, Kojima S, Yoshiie T and Kiritani M 1991 *J. Nucl. Mater.* **179-181** 901
[16] Robinson M T and Torrens I M 1974 *Phys. Rev.* B **9** 5008
[17] Kinchin G H and Pease R S 1955 *Rep. Prog. Phys.* **18** 1
[18] Greenwood G R and Smither P K 1985 "SPECTER:Neutron Damage Calculations for Materials Irradiation", ANL/FPP/TM-197
[19] Ishida I, Yoshiie T, Sasaki S, Iwase A, Iwata T and Kiritani M 1988 *J. Nucl. Mater.* **155-157** 417
[20] Kiritani M 1989 *J. Nucl. Mater.* **169** 89
[21] Garner F A 1993 *J. Nucl. Mater.* **205** 98
[22] Mansur L K and Coghlan W A 1983 *J. Nucl. Mater.* **119** 1
[23] Tanaka M P, Hamada S, Hishinuma A and Maziasz P J 1988 *J. Nucl. Mater.* **155–157** 801

Journal of Physics: Conference Series **674** (2016) 012003 doi:10.1088/1742-6596/674/1/012003

Effect of Cu content on the defect evolution in Fe-Cu alloys investigated by PALS

X Z Cao[1], X Lai[1,2], G D Cheng[1], S X Jin[1], X Zhang[2], P Zhang[1], B Y Wang[1,*]

[1]Key Laboratory of Nuclear Radiation and Nuclear Energy Technology, Institute of High Energy Physics, CAS, Beijing 100049, China
[2]Key Laboratory of High Energy Density Physics and Technology of Ministry of Education, Sichuan University, Chengdu 610064, China

E-mail: wangboy@ihep.ac.cn

Abstract. The effect of Cu content on the evolution of defects in Fe-x%Cu alloys (x= 0.15, 0.3, and 0.6) were investigated using Positron Annihilation Lifetime Spectroscopy (PALS). The vacancy-type and Cu-vacancy complexes defects were respectively produced by quenching from 1173 K and cold-rolled deformation followed by isochronal annealing. The PALS results with isochronal annealing showed that the temperature of defects recovery increased in deformed Fe-Cu alloys with Cu content. The increment of Cu content also restrained the migration of vacancies in as-quenched Fe-Cu alloys.

1. Introduction

Fe-Cu model alloy has been always regarded as one of the most suitable systems for studying the Cu precipitates and the evolution of micro defects in the reactor pressure vessel (RPV) steels [1,2]. Extensive studies have showed that the increment of the ductile-to-brittle transition temperature (DBTT) was due to the tiny Cu precipitates in RPV steels [1,3,4]. Since the influence of small amount Cu element on DBTT of steel used for RPV steels is a serious problem, the content of the Cu impurity must be limited strictly in commercial reactor structural materials. It is well known that Cu atoms have a very low solubility in Fe-base alloys and the Cu precipitates formed easily in Cu-bearing steels during deformation, irradiation and aging. Formation of Cu precipitates, evolution of defects and the interaction between Cu precipitates and defects have been widely investigated [5,6]. Formation of Cu-vacancy clusters and the deformation-induced vacancies enhanced the diffusion of Cu atoms were also investigated using positron annihilation spectroscopy in rolling deformed Fe-Cu alloys with isochronal treatment [7]. In addition, Nagai et al. investigated the interaction of irradiation-induced vacancies and Cu aggregations in binary Fe-Cu alloys, they found that irradiation could lead to a vacancy-solute complex which would speed up the precipitation [8]. Although the Fe-Cu binary alloys have been extensively studied, the effect of Cu content on the evolution of defects along with the interaction between Cu atoms and vacancies in quenched and deformed Fe-Cu alloys are still poorly understood. However, the mechanism of the interaction of Cu atoms with defects plays a critical role in the properties of Fe-Cu alloys.

It is difficult to probe experimentally using conventional detection means such as Transmission Electron Microscopy (TEM) as many of these defects are on atomic scale. Positron annihilation techniques (PAT) have been used to investigate tiny Cu precipitates, defects and their complexes in

Fe-Cu alloys [9]. Importantly, PALS has been widely applied as one of the few techniques to detect the evolution of vacancy-type defects sensitively and selectively in metals [10-12].

In order to clarify the effect of Cu content on the evolution of defects and the interaction between Cu atoms and vacancies, PALS was performed on three sorts of Fe-x%Cu alloys (x= 0.15, 0.3, and 0.6) treated with different ways. Deformation and quenching were used to produce defects in Fe-x%Cu alloys. Isochronal annealing was a common method to study the evolution and recovery of defects.

2. Experiments

As specimens, Fe-x%Cu alloys (x= 0.15, 0.3, and 0.6) were arc melted from Fe (99.99 % purity) and Cu (99.999 % purity) in vacuum using a high-frequency induction furnace. After melting, solution treatment of the samples consisted of heating at 1073 K for 24 h, and then well-annealed at 1173 K for 0.5 h in vacuum, followed by quenching into ice water to introduce vacancies [13]. The other series of specimens were cold-rolling deformed to 30% in order to induce the complex defects after quenching. After introduction of defects by quenching and deformation, the quenching specimens were isochronally annealed for 30 min in a vacuum of 10^{-5} Pa from 373 K to 673 K; and the deformed specimens was annealed from 373 K to 973 K. PALS analysis was carried out to characterize the defects in Fe-Cu alloys using a fast-slow coincident ORTEC system with a time resolution of 200 ps (full width at half maximum). The ^{22}Na positron source was placed between the two pieces of samples, and then the "sample-source-sample sandwich" was placed between the two BaF_2 detectors to acquire the lifetime spectra. Each spectrum was accumulated to a total of 2×10^6 counts, which aimed to reduce the statistical error in the calculation of lifetimes. The positron lifetime spectra were analyzed by the LT-9 software [7]. Positron annihilation lifetime and intensity could reflect the defect size and density information.

3. Results and Discussion

The results of isochronal annealing for as-quenched Fe-x%Cu alloys with 30% deformation obtained by PALS are shown in Figure 1. τ_m is the mean lifetime which reflects the total amount of vacancy-type defects in alloys. The long lifetime τ_2 corresponds to the size of vacancy clusters [14]. I_2 is the intensity of τ_2 which correlate the defect density. The short lifetime τ_1 may be contributed by positron lifetime of interstitial atoms and other defects, such as dislocation or bulk material [15]. It is seen that the value of τ_m in Fe-x%Cu alloys after 30% deformation were about 150 ps, and the mean lifetime of a perfect lattice in Fe was about 106 ps [16]. The τ_m is distinctly longer than the lifetime of Fe bulk, which indicated that a significant amount of defects were induced by deformation. The long lifetime τ_2 remained almost constant value after isochronal annealing at the temperature region 373 K-523 K, while the intensity of τ_2 increased gradually. The reason may be that a small amount of deformation-induced dislocations could trap vacancies, which could be released gradually with the increment of annealing temperature and finally formed to vacancy clusters or annihilated at the vacancy sinks, as the increasing of long lifetime (τ_2) intensity (I_2). As shown in Figure 1, the long lifetime τ_2 for Fe-x%Cu alloys all increased to around 200 ps after isochronal annealing at 573 K, suggesting the vacancies aggregated and formed the vacancy clusters. It was worth noting that the annealing behavior of the long lifetime in Fe-0.15%Cu was slightly different from that in Fe-0.3%Cu and Fe-0.6%Cu from 523 K to 623 K, the long lifetime in Fe-0.15%Cu increased to around 223 ps and remained stable, while the τ_2 of Fe-0.3%Cu and Fe-0.6%Cu decreased to around 160 ps subsequently. It indicated that the behavior of defects from 523 K to 623 K related to Cu content, namely, the increment of Cu content may promote the occupation of vacancies by Cu atoms and benefit for the formation of Cu-vacancy complexes further. Thus the increase of Cu content may enhance the separation of vacancy clusters in Fe-Cu alloys. On the other hand, when the mean lifetime decreased to the lifetime of a perfect lattice in pure Fe (~106 ps), this marked the complete recovery of defects [16]. The results of the mean lifetime in Figure 1 show that the perfect recovery temperature was 923 K in Fe-x%Cu (x= 0.15, 0.3) and 973 K in Fe-0.6%Cu. Cu-vacancy complexes had gradually transformed to Cu precipitates around 823 K after annealing and Cu precipitates may be formed on dislocations in matrix. The recovery process of the precipitates is strongly affected by the Cu content, while the recovery temperature corresponds to the solubility limit. As a result, it could be concluded that Fe-0.6%Cu had a higher recovery temperature compared to Fe-x%Cu (x=0.15, 0.3) because of the solubility limit, namely Cu.

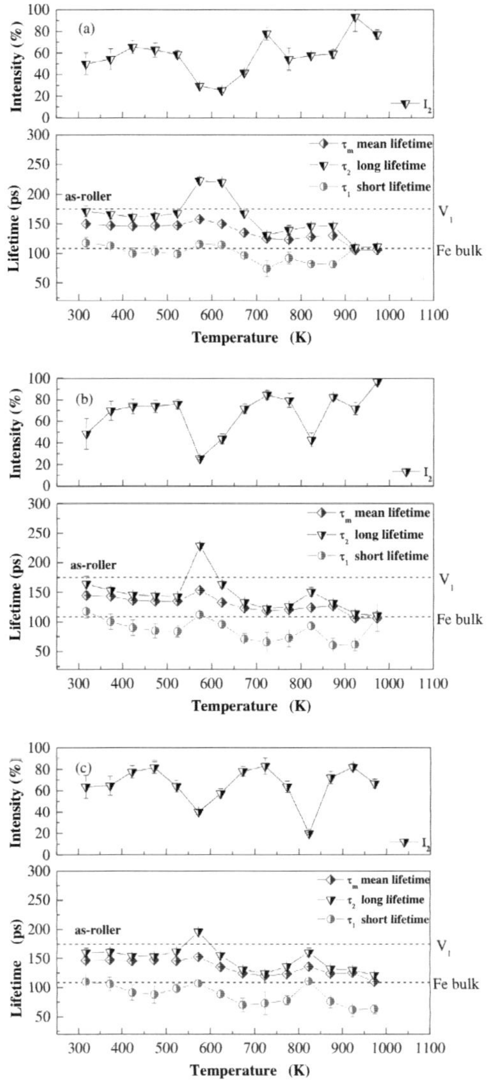

Figure 1. PALS for quenched Fe-x%Cu alloys with 30% deformation. (a) Fe-0.15%Cu (b) Fe-0.3%Cu (c) Fe-0.6%Cu. The calculated values for bulk and V_1 in pure Fe are shown by horizontal dashed lines.

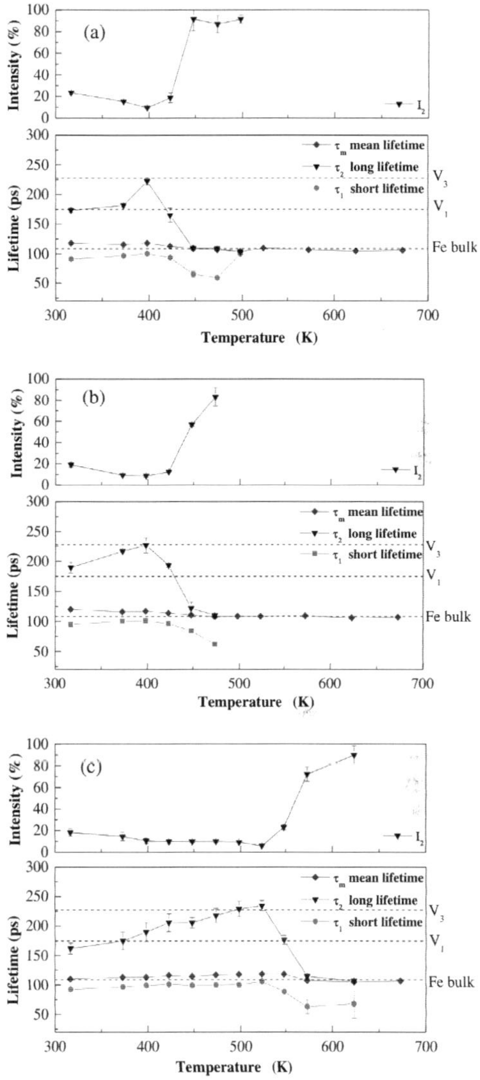

Figure 2. PALS for only quenched Fe-x%Cu alloys. (a) Fe-0.15%Cu (b) Fe-0.3%Cu (c) Fe-0.6%Cu. The calculated values for bulk, V_1 and V_3 in pure Fe are shown by horizontal dashed lines

In order to further investigate the influence of Cu content on the evolution of vacancy-type defects in Fe-Cu alloys and rule out the effect of deformation-induced Cu-vacancy complexes on the behaviour of vacancy-type defects, the only quenched specimens were investigated by PALS. As shown in Figure 2, the complete recovery temperature of Fe-0.15%Cu, Fe-0.3%Cu and Fe-0.6%Cu after isochronal annealing is 448 K, 473 K and 623 K, respectively. This phenomenon also demonstrated that the complete recovery temperature of defects increased with the Cu content because of the solubility limit. The complete temperatures for deformed Fe-x%Cu alloy in Figure 1 were distinctly larger than those for quenched samples in Figure 2. Therefore, Cu-vacancy complexes induced by deformation pushed the complete recovery temperature higher indirectly. As shown in Figure 2, the long lifetime in Fe-0.15%Cu and Fe-0.3%Cu increased to around 227 ps at 398 K after annealing, which meant single vacancies migrated and formed clusters. However, the corresponding temperature in Fe-0.6%Cu was 523 K. This also demonstrated that Cu content had an effect on the migration of vacancies in as-quenched Fe-Cu alloys. To be precise, Cu atoms may restrain the migration of vacancies.

4. Conclusion

The effect of Cu content in Fe-x%Cu (x= 0.15, 0.3, and 0.6) on the evolution of defects were investigated systematically by PALS. In summary, according to the results of PALS, the variation of the temperature where three sorts of specimens completely recovered revealed that the complete recovery temperature presented a significant increase with the increase of Cu content. It could be assumed that Cu precipitates transformed from Cu-vacancy complexes induced by deformation and subsequent isochronal annealing pushed the complete recovery temperature higher because of the solubility limit. In addition, Cu content had an effect on the migration of vacancies in as-quenched Fe-Cu alloys. To be precise, Cu atoms may restrain the migration of vacancies.

Acknowledgement

This work is partially supported by the National Natural Science Foundation (Grant Nos. 91026006, 91226103).

References

[1] Phythian W J and English C A, 1993 *J. Nucl. Mater.* **205** 162
[2] Odette G R and Lucas G E 1998 *Radiat. Eff. Defects Solids* **144** 189
[3] Cao X Z, Xu Q, Sato K and Yoshiie T 2009 *Phys. Status Solidi* C **6** 2355-2358
[4] Nagai Y, Tang Z, Hassegawa M, Kanai T and Saneyasu M 2001 *Phys. Rev.* B **63** 134110
[5] Cao X Z, Zhang P, Xu Q, Sato K, Tsuchida H, Cheng G D, Wu H B, Jiang X P, Yu R S, Wang B Y and Wei L 2013 *J. Phys. Conf. Ser.* **443** 012017
[6] Xu Q and Yoshiie T 2011 *Philos. Mag.* **91** 3716
[7] Onitsuka T, Takenaka M, Kuramoto E, Nagai Y and Hasegawa M 2001 *Phys. Rev.* B **65** 012204
[8] Nagai Y, Tang Z, Ohkubo H, Takadate K and Hasegawa M 2003 *Radiat Phys. Chem.* **68** 381
[9] Nagai Y, Takadate K, Tang Z, Ohkubo H, Sunaga H, Takizawa H and Hasegawa M 2003 *Phys. Rev.* B **67** 224202
[10] Ohkubo H, Sugiyama S, Fukuzato K, Takenaka M, Tsukuda N and Kuramoto E 2000 *J. Nucl. Mater.* **283** 858
[11] Nagai Y, Hasegawa M, Tang Z, Hempel A, Yubuta K, Shimamura T, Kawazoe Y, Kawai A and Kano F 2000 *Phys. Rev.* B **61** 6574
[12] Yoshiie T, Xu Q, Satoh Y, Ohkubo H and Kiritani M 2000 *J. Nucl. Mater.* **283** 229
[13] Yoshiie T, Cao X Z, Xu Q, Sato K and Troev T D 2009 *Phys. Status Solidi* C **6** 2333
[14] Ishizaki T, Xu Q, Yoshiie T and Nagata S 2004 *Mater. Trans.* **45** 9
[15] Wu H B, Cao X Z, Cheng G D, Wu J P, Yang J, Zhang P, Li Z X, Rahman A M S, Yu R S and Wang B Y 2013 *Phys. Status Solidi* A **210** 1758
[16] Ohkubo H, Tang Z, Nagai Y, Hasegawa M, Tawara T and Kiritani M 2003 *Mat. Sci. Eng.* A **350** 95

Early stages of precipitation in Mg-RE alloys studied by positron annihilation spectroscopy

M Vlček[1], J Čížek[1], O Melikhova[1], P Hruška[1], I Procházka[1], M Vlach[1], I Stulíková[1], B Smola[1]
[1]Charles University in Prague, Faculty of Mathematics and Physics, V Holešovičkách 2, 18000 Praha 8, Czech Republic

E-mail: marian.vlcek@gmail.com

Abstract. Magnesium alloys with rare earth (RE) elements are promising structural materials exhibiting favourable mechanical properties at elevated temperatures. However, the processes occurring during early stages of precipitation in these alloys are still not completely understood. In this work positron lifetime spectroscopy combined with coincidence Doppler broadening was employed for investigation of early stages of precipitation in Mg-RE alloys. Presence of quenched-in vacancy clusters was observed after solution treatment of studied alloys. These quenched-in vacancy clusters are bound to RE solutes and thereby stabilized at room temperature. During natural aging, RE clusters are formed by vacancy-assisted long-range diffusion. In addition, hardness of studied materials increases and quenched-in vacancy clusters are annealed out during the course of natural aging. Simple model was developed to describe hardening during natural aging.

1. Introduction

Mg-based alloys with rare earth (RE) alloying elements, e.g. Gd and Tb, represent promising light hardenable alloys with a high creep resistance at elevated temperatures [1]. Since the solubility limit of RE alloying elements in Mg strongly decreases at low temperatures, supersaturated solid solution of RE solutes in α-Mg matrix can be formed by fast quenching from elevated temperatures. During subsequent heating the supersaturated solid solution decomposes by formation of a sequence of metastable phases which may cause significant hardening of the alloy. For example the supersaturated solid solution of Gd in Mg decomposes in the sequence [2]:

$$\alpha'(\text{hcp}) \rightarrow \beta''(\text{DO}_{19}\text{ clusters}) \rightarrow \beta'(\text{c-bco}) \rightarrow \beta\ (\text{fcc}, \text{Mg}_5\text{Gd type}). \qquad (1)$$

The β'' particles are coherent with the α-Mg matrix while for the β' and β precipitates the coherency with the α-Mg matrix is partially lost. Although the precipitation sequences are nowadays well documented for various Mg-RE systems, there is still a lack of information about very early stages of decomposition leading eventually to formation of coherent metastable particles usually with DO_{19} structure.

Natural aging is a process where solute atoms and vacancies cluster at room temperature in materials quenched from high annealing temperature. It is well known effect and was thoroughly investigated in Al-based alloys [3, 4, 5, 6]. Significant strengthening occurs because solute clusters developed during natural aging hinder movement of dislocations. Contrary to Al-based alloys, natural aging of Mg-based alloys is not common. So far, natural aging was observed in Mg-Zn-based alloys [7] and, recently, in our work in Mg-Gd and Mg-Tb alloys [8]. In this work positron lifetime spectroscopy combined with coincidence Doppler broadening was employed for investigation of early stages of precipitation in Mg-RE alloys.

2. Experimental details

Binary Mg-Gd and Mg-Tb were produced by squeeze casting under a protective atmosphere (Ar + 1% SF_6). Chemical composition of studied alloys is shown in table 1.

Table 1. Composition of studied alloys.

	Gd		Tb		Mg
	wt.%	**at.%**	**wt.%**	**at.%**	
Mg4Gd	4.48	0.72	-	-	balance
Mg9Gd	9.24	1.55	-	-	balance
Mg15Gd	14.58	2.57	-	-	balance
Mg13Tb	-	-	13.39	2.31	balance

The as-cast Mg-Gd and Mg-Tb alloys were subjected to solution treatment at 500 and 530°C, respectively, for 6 hours finished by quenching into water at room temperature. Samples were subsequently naturally aged at ambient temperature.

Positron lifetime spectroscopy measurements were performed using a digital positron lifetime spectrometer [9] with excellent time resolution of 145 ps (FWHM of the resolution function obtained from fitting of positron lifetime spectrum of a well-annealed Mg reference sample). At least 10^7 annihilation events were accumulated in each spectrum. Source
contribution consisting of two components with lifetimes 368 ps and 1.5 ns with relative intensities 8% and 1%, respectively, which originated from positrons annihilated in the $^{22}Na_2CO_3$ positron source deposited on a 2 μm thick mylar foil, was substracted from the spectra.

Coincidence Doppler broadening (CDB) measurements were carried out using a digital spectrometer [10] equipped with two HPGe detectors with the resolution of 0.9 keV at 511 keV (FWHM). The total statistics accumulated in each two-dimensional CDB spectrum was at least 10^8 positron annihilation events in each spectrum.

The Vickers microhardness (HV) testing was carried out using a STRUERS Duramin 300 hardness tester with a load of 100 g applied for 10 s.

3. Results

Time development of hardness of solution treated samples aged at ambient temperature is shown in figure 1a. Hardness of all studied alloys clearly increases with increasing aging time. The time dependence of microhardness plotted in the logarithmic time scale exhibits characteristic S-shaped curve typical for natural aging.

The hardness of alloy aged for time t can be expressed as:

$$HV(t) = HV_0 + \Delta HV(t), \tag{2}$$

where $HV_0 = 30 \pm 2$ is the hardness of pure Mg matrix and $\Delta HV(t)$ denotes hardening caused by solute elements:

$$\Delta HV(t) = (HV_{ss}^2 + HV_{cl}^2)^{1/2}. \tag{3}$$

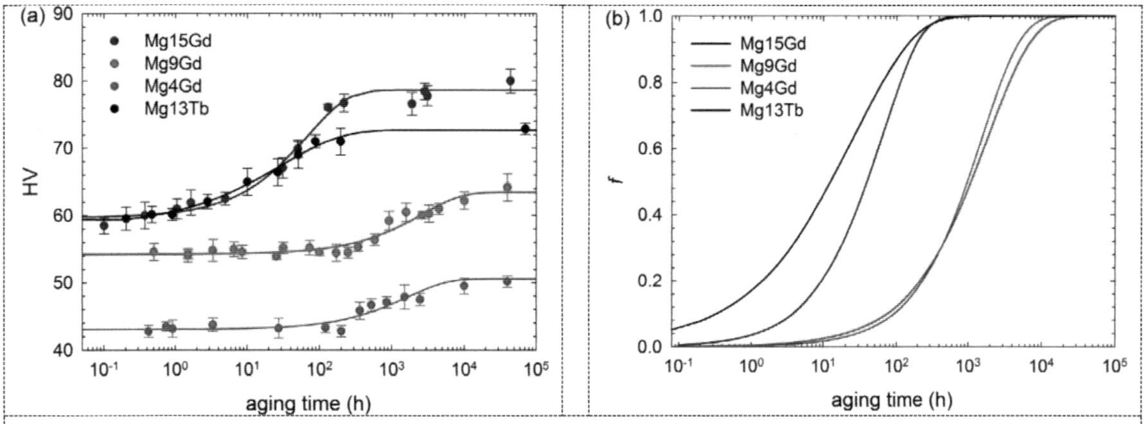

Figure 1. (a) Comparison of microhardness evolution during natural aging of studied alloys. (b) Time development of relative volume fraction of clusters obtained from fitting.

The symbols HV_{ss} and HV_{cl} denote contribution to the hardening caused by solid solution hardening and by clusters of alloying elements, respectively. Let f denote the relative fraction of solute atoms present in clusters. The hardening caused by clusters HV_{cl} is then proportional to $f^{1/2}$ while the solid solution hardening HV_{ss} is proportional to $(1-f)^{2/3}$ [3]. Hence, the evolution of the hardness during aging can be written as:

$$HV(t) = HV_0 + \left[\left(h_{ss}c^{2/3}(1-f)^{2/3}\right)^2 + \left(h_{cl}\left(\frac{c}{N_{cl}}\right)^{1/2} f^{1/2}\right)^2\right]^{1/2}, \tag{4}$$

where c is the total concentration of alloying element (Gd or Tb) in at.%, h_{ss} and h_{cl} are hardening coefficients for solid solution hardening and hardening by clusters, respectively, and N_{cl} is the average number of atoms per cluster.

Table 2. Fitted parameters of hardening model.

	h_{ss}	$h_{cl}/(N_{cl})^{1/2}$	k (s^{-1})	n
Mg4Gd	350(10)	243(8)	1.8(7)×10^{-7}	0.8(3)
Mg9Gd	390(5)	269(5)	1.5(4)×10^{-7}	0.7(1)
Mg15Gd	342(8)	301(3)	4.5(7)×10^{-6}	0.8(1)
Mg13Tb	360(10)	286(4)	1.1(2)×10^{-5}	0.52(8)

The clusters in aged alloys develop by nucleation and subsequent growth. Since the studied alloys are coarse grained and exhibit low dislocation density, it is reasonable to assume that heterogeneous nucleation in limited number of suitable nucleation sites takes place. These sites are used at the early stage of clustering and so called site saturation occurs. Under these assumptions, the relative volume fraction of clusters is given by expression:

$$f = 1 - \exp[-(k(T)t)^n], \tag{5}$$

where $k(T)$ is the kinetic rate at temperature T and n is the kinetic exponent [3]. The quantity f is also identical to the relative fraction of solute atoms which precipitated from the solid solution and formed clusters. Hence, when $f = 0$ all solute atoms are in solid solution and no clusters are present in the sample, while the clusters are fully developed and no solutes remain in solid solution when $f = 1$.

Results obtained by fitting the measured data with proposed model of hardening are shown in table 2. The relative volume fraction of clusters f obtained from fitting is shown in figure 1b. As an example, fits of data measured on Mg15Gd and Mg13Tb alloys showing individual contributions HV_0, HV_{ss} and HV_{cl} to the total hardness are shown in figures 2a and 2b, respectively.

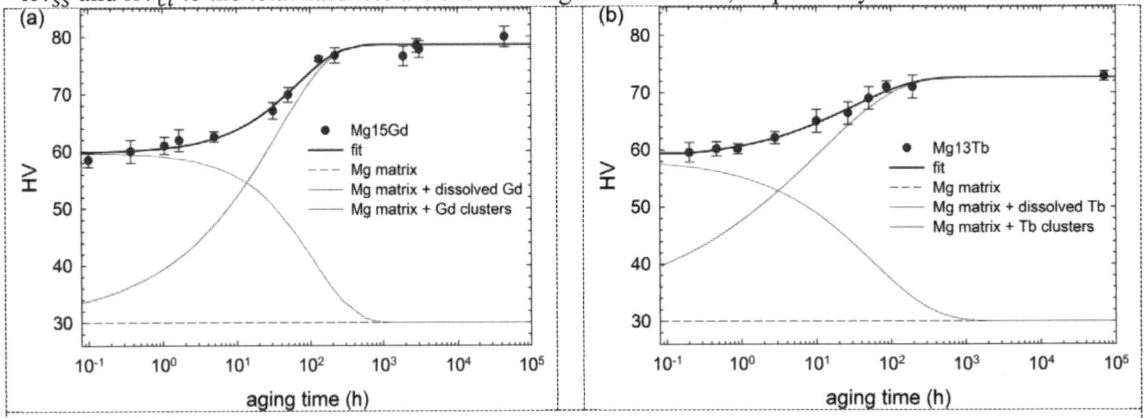

Figure 2. Development of microhardness and fit by the theoretical model described in the text: (a) Mg15Gd, (b) Mg13Tb.

Results obtained by positron lifetime spectroscopy are summarized in table 3. All quenched alloys exhibit two component positron lifetime spectra. The first component comes from free positrons while the second component represents a contribution of positrons trapped at defects. The lifetime of the second component $\tau_2 \approx 300$ ps agrees well with the theoretically calculated lifetime for a monovacancy in Mg [11]. However, experimental results suggest that positron lifetime for a monovacancy in Mg is ≈ 255 ps [12]. Therefore, positron lifetime measurements testify to presence of quenched-in vacancies or small vacancy clusters in the quenched alloys.

On the other hand, the alloys aged at ambient temperature for 2 months exhibit single component spectra with lifetime $\tau_1 \approx 225$ ps which agrees well with the bulk positron lifetime in Mg [11]. Hence, they are defect free within the sensitivity of positron lifetime spectroscopy.

Table 3. Results of positron lifetime measurements.

	quenched alloys				aged alloys (2 months)	
	τ_1 (ps)	I_1 (%)	τ_2 (ps)	I_2 (%)	τ_1 (ps)	I_1 (%)
Mg4Gd	219.7(5)	92.5(5)	290(10)	7.5(5)	225.5(2)	100
Mg9Gd	218.6(5)	88.4(5)	300(5)	11.6(5)	225.2(2)	100
Mg15Gd	214.6(7)	80.5(6)	295(5)	19.5(6)	225.4(2)	100
Mg13Tb	214.0(8)	84.3(7)	280(15)	15.7(7)	225.5(2)	100

Results of coincidence Doppler broadening measurements performed on Mg15Gd and Mg13Tb alloys are shown in figures 3a and 3 b, respectively. Ratio curves of quenched alloys show peak located at $p \approx 8 \times 10^{-3}\ m_0c$ which testifies to enhanced concentration of Gd or Tb around quenched-in vacancy clusters. Amplitude of this peak significantly decreases after aging due to disappearance of quenched-in vacancy clusters.

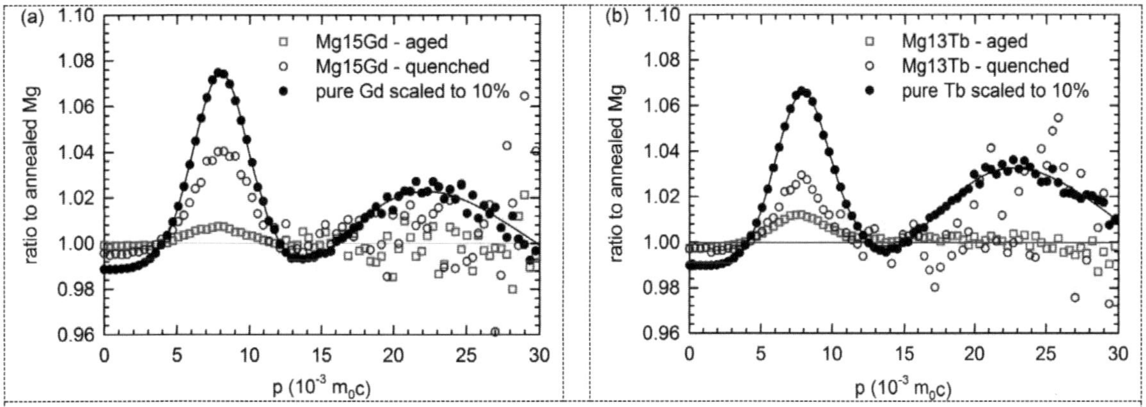

Figure 3. Coincidence Doppler broadening ratio curves: (a) Mg15Gd, (b) Mg13Tb.

Under the assumption that vacancy-Gd (or Tb) complexes consist of Mg monovacancy and a solute atom, concentration of quenched-in vacancy-Gd (or Tb) complexes can be calculated from positron lifetime data according to the simple trapping model [13]:

$$c_{pairs} = \frac{1}{\nu_v} I_2 \left(\frac{1}{\tau_1} - \frac{1}{\tau_2}\right), \tag{6}$$

where $\nu_v = 1.1 \times 10^{13}$ s^{-1} is the specific positron trapping rate for vacancy in Mg [14].

Note that the specific positron trapping rate for a small cluster composed of N vacancies is $N\nu_v$. Therefore, trapping rate to vacancies of fixed concentration is the same independently whether they exist as isolated monovacancies or form small vacancy clusters providing that positron trapping is in transition limited regime [15], i.e. the mutual distance among vacancy clusters is significantly smaller than the mean positron diffusion length.

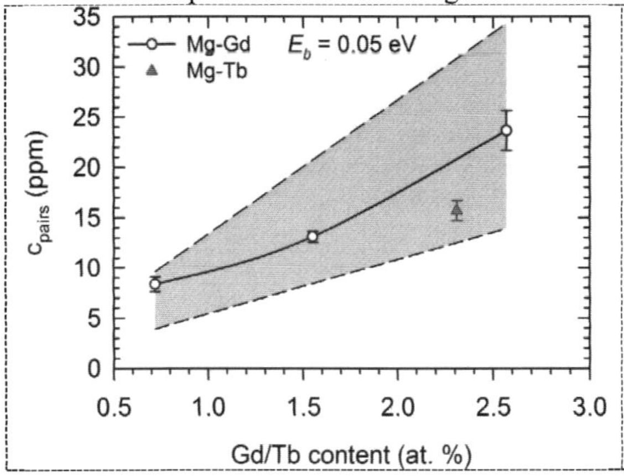

Figure 4. The concentration quenched-in vacancy-Gd (open circles) and vacancy-Tb (full triangle) pairs measured by positron lifetime spectroscopy. Shaded region shows the band for equilibrium concentration of vacancy-solute pairs at 500°C calculated by equation (7). Binding energy E_b between vacancy and Gd or Tb was assumed to be 0.05 eV.

Values calculated according to equation 6 can be compared to the equilibrium concentration of vacancy-Gd (or Tb) pairs at solution treatment temperature T given by expression:

$$c^*_{pairs} = c \exp\left(\frac{S_{v,f}}{k_B}\right) \exp\left(-\frac{E_{v,f} - E_b}{k_B T}\right), \tag{7}$$

where k_B is the Boltzmann constant, $S_{v,f} \approx 2k_B$ is the vacancy formation entropy [16]. The values of vacancy formation enthalpy $E_{v,f}$ reported in literature fall into the range 0.79-0.85 eV [12,17,18]. E_b is the binding energy between vacancy and Gd or Tb. The latter quantity is not known, but putting $E_b = 0.05$ eV, i.e. assuming small attractive interaction between vacancy and Gd or Tb atom, the equilibrium concentration of vacancy-Gd(Tb) pairs given by equation (7) falls into the shaded area in figure 4 which is in good agreement with experimental data.

4. Discussion

All studied Mg-RE alloys exhibit natural aging and the magnitude of hardening during aging as well as the concentration of quenched-in vacancy clusters increase with increasing concentration of RE alloying elements. It was found that early stages of precipitation are controlled by quenched-in vacancies. Positive binding energy between vacancy and solute atoms facilitates creation of solute-vacancy complexes during solution treatment at elevated temperature. However, since the binding energy between vacancy and Gd or Tb is comparable to the thermal energy most vacancies are not bound to Gd or Tb solutes during the solution treatment. In quenched samples free vacancies are quickly annealed out at room temperature but they can also bind to RE solutes. Vacancies associated with RE solutes are more stable and facilitate diffusivity of the solute RE atoms. Hence, the early stages of precipitation are strongly affected by the concentration of quenched-in vacancies bound to RE solutes. Agglomeration of solutes to small clusters proceeds by vacancy-assisted long-range diffusion. Solute clusters with defected structure containing vacancies are formed in the initial stage and further develop into precursors of the coherent DO_{19} precipitates.

Ab-initio theoretical calculations of positron lifetime for vacancy clusters with different sizes were performed to estimate the number of vacancies in vacancy-Gd (or Tb) complexes. Positron lifetimes were calculated within the so-called standard scheme [19] employing the atomic superposition method [20]. The electron-positron correlation was treated using the local density approximation according to the parametrization by Boroński and Nieminen [21]. The calculations were performed in 512 Mg atom based supercells. Vacancies and vacancy clusters were modeled simply by removing the corresponding number of atoms from the supercell.

Results of these calculations are presented in figure 5. Measured positron lifetime of defect component in quenched samples agrees well with the lifetime of ≈ 300 ps calculated for a Mg monovacancy. Note that similar lifetime was calculated for Mg monovacancy in reference [11].

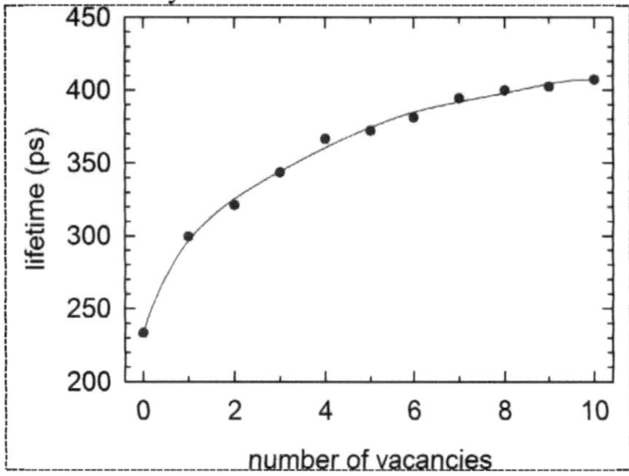

Figure 5. Calculated dependence of positron lifetime on the number of Mg vacancies in vacancy cluster. The solid line is just to guide the eyes.

On the other hand, the experimental lifetime value of ≈ 255 ps was determined by in-situ positron lifetime study of Mg samples annealed in the temperature range 0-600°C through fitting of the temperature dependence of the mean positron lifetime [12]. Low temperature measurements of quenched and electron irradiated Mg samples performed in reference [12] revealed defect component with positron lifetime of (350 ± 20) ps. Comparison with theoretical calculations suggests that in quenched and irradiated samples positron trapping occurs at complex small vacancy clusters consisting of 2-3 vacancies.

Nevertheless, the large difference between experimental and theoretical lifetimes for Mg monovacancy is very remarkable. In addition, the experimental value ≈ 255 ps determined in reference [12] is practically the same as the positron lifetime determined experimentally for dislocations in Mg which is ≈ 250 ps [22, 23, 24]. Therefore, further investigations are needed to definitively resolve these discrepancies and firmly assign the positron lifetime for vacancy in Mg. Hence, at the present stage of knowledge we can conclude that quenched Mg-Gd and Mg-Tb alloys studied here contain either monovacancies or small vacancy clusters (consisting of 2-3 vacancies) bound to Gd or Tb solutes.

The kinetics of natural aging is not the same for all studied alloys. The Mg15Gd and Mg13Tb alloys show significantly shorter incubation period of hardening than Mg4Gd and Mg9Gd alloys. This is most likely caused by higher concentration of alloying elements and thereby also higher concentration of quenched-in vacancies bound to RE solutes. However, further studies are required to elucidate the factors influencing the kinetics of natural aging in Mg-RE alloys.

5. Conclusions

Natural aging of Mg-Gd and Mg-Tb alloys associated with clustering of solute atoms was observed in this work. Theoretical model of hardening during natural aging showing good agreement with experimental data was developed.

Presence of quenched-in vacancy/vacancy cluster-Gd (or Tb) complexes in solution treated alloys was confirmed by positron annihilation spectroscopy. Quenched-in vacancies or vacancy clusters composed of up to three vacancies facilitate diffusion of Gd or Tb atoms in Mg matrix and disappear when the development of solute clusters is finished.

6. Acknowledgements

The research leading to these results has received funding from the People Programme (Marie Curie Actions) of the European Union's Seventh Framework Programme FP7/2007-2013/ under REA grant agreement N° 289163.

One of the authors (P. Hruška) acknowledges support by the Grant Agency of the Charles University (project no. 324015).

7. References

[1] Mordike B 2002 *Mat. Sci. Eng. A* **324** 103
[2] Vostrý P, Stulíková I, Smola B, von Buch F and Mordike B L 1999 *Phys. Status Solidi A* **175** 491
[3] Esmaeili S, Lloyd D J and Poole W J 2003 *Acta Materialia* **51** 3467
[4] Stulíková I, Faltus J and Smola B 2003 *Kovove Mater.* **45** 85
[5] Klobes B, Staab T E M, Haaks M, Maier K and Wieler I 2008 *Phys. Status Solidi RRL* **2** 224
[6] Banhart J, Lay M D H, Chang C S T and Hill J *2011 Phys. Rev. B* **83** 014101
[7] Buha J 2008 *J. Mater. Sci.* **43** 1120
[8] Čížek J, Smola B, Stulíková I, Hruška P, Vlach M, Vlček M, Melikhova O and Procházka I 2012 *Phys. Status Solidi A* **209** 2135
[9] Bečvář F, Čížek J, Procházka I and Janotová J 2005 *Nucl. Instr. Meth. Phys. Res. A* **539** 372
[10] Čížek J, Vlček M and Procházka I 2010 *Nucl. Instrum. Methods A* **623** 982
[11] Robles J M C, Ogando E and Plazaola F 2007 *J. Phys.: Condens. Matter* **19** 176222
[12] Hautojärvi P, Johansson J, Vehanen A and Yli-Kauppila J 1982 *Appl. Phys. A* **27** 49
[13] West R 1979 Positrons in Solids ed Hautojärvi P (Berlin: Springer-Verlag) p 89
[14] Hood G M 1982 *Phys. Rev. B* **26** 1036
[15] Dupasquier A, Romero R and Somoza A 1993 *Phys. Rev. B* **48** 9235
[16] Wollenberger H 1983 *Physical Metalurgy* vol 2 (Amsterdam: North-Holland) p 1146
[17] Tzanetakis P, Hillairet J and Revel G 1976 *Phys. Status Solidi B* **75** 433
[18] Tzanetakis P 1978 *Thése* (Université de Grenoble)

[19] Puska M J and Nieminen R M 1994 *Rev. Mod. Phys.* **66** 841
[20] Puska M J and Nieminen R M 1983 J. *Phys. F: Met. Phys.* **13** 333
[21] Boroński E and Nieminen R M 1986 *Phys. Rev. B* **34** 3820
[22] Čížek J, Procházka I, Smola B, Stulíková I, Kužel R, Matěj Z and Cherkaska V 2006 *Phys. Status Solidi A* **203** 466
[23] del Río J, Gómez C and Ruano M 2012 *Philos. Mag.* **92** 535
[24] Dryzek J and Dryzek E 2011 *Magnesium Alloys - Corrosion and Surface Treatments* (Rijeka: InTech) p 289

Journal of Physics: Conference Series **674** (2016) 012005 doi:10.1088/1742-6596/674/1/012005

Effects in Mg-Zn-based alloys strengthened by quasicrystalline phase

M Vlček[1], J Čížek[1], F Lukáč[1], O Melikhova[1], P Hruška[1], I Procházka[1], M Vlach[1], I Stulíková[1], B Smola[1], A Jäger[2]

[1]Charles University in Prague, Faculty of Mathematics and Physics, V Holešovičkách 2, 18000 Praha 8, Czech Republic

[1]Institute of Physics, Academy of Sciences of the Czech Republic, Na Slovance 2, 18221 Praha 8, Czech Republic

E-mail: marian.vlcek@gmail.com

Abstract. Magnesium Mg-based alloys are promising lightweight structural materials for automotive, aerospace and biomedical applications. Recently Mg-Zn-Y system attracted a great attention due to a stable icosahedral phase (I-phase) with quasicrystalline structure which is formed in these alloys. Positron lifetime spectroscopy and *in situ* synchrotron X-ray diffraction were used to study thermal stability of I-phase and precipitation effects in Mg-Zn-Y and Mg-Zn-Al alloys. All alloys containing quasicrystalline I-phase exhibit misfit defects characterized by positron lifetime of ~ 300 ps. These defects are associated with the interfaces between I-phase particles and Mg matrix. The quasicrystalline I-phase particles were found to be stable up to temperatures as high as ~ 370°C. The W-phase is more stable and melts at ~ 420°C. Concentration of defects associated with I-phase decreases after annealing at temperatures above ~ 300°C.

1. Introduction

Mg-Zn-Y system attracts attention due to presence of icosahedral phase (I-phase) with quasicrystalline structure [1]. Quasicrystals exhibit structure with lack of translation symmetry and posses unique combination of physical properties: high hardness, high corrosion resistance and low surface energy [2]. Although they are not directly applicable for structural applications due to their brittleness, they could be used as strengthening agents in Mg alloys. Moreover, I-phase has strong interfacial bonding to the Mg matrix, low friction coefficient and low surface energy which is beneficial for ductility and deformability.

Positron annihilation spectroscopy revealed existence of misfit defects associated with the interfaces between I-phase particles and α-Mg matrix [3]. Thermal stability of I-phase and precipitation effects in Mg-Zn-Y and Mg-Zn-Al alloys were studied in this work using positron lifetime spectroscopy combined with *in situ* synchrotron X-ray diffraction (XRD). Development of mechanical properties was monitored by microhardness testing.

2. Experimental details

Three groups of alloys were studied in this work: (i) Mg-Y-Nd-Zr (WE43) alloy, (ii) WE43 alloy modified by addition of various amount of Zn, and (iii) Mg-Zn-Al alloys. Both WE43 modified by Zn and Mg-Zn-Al alloys contain quasicrystalline I-phase. The WE43 alloy without Zn does not contain quasicrystalline phase and was used as a reference material. All alloys were produced by squeeze casting under a protective atmosphere (Ar + 1% SF_6). Chemical compositions of studied alloys are shown in tables 1 and 2.

Table 1. Compositions of Mg-Zn-Y-based alloys in wt.%.

	Zn	Y	Nd	Zr	Gd	Mg
WE43	-	2.95	2.48	0.30	0.15	balance
WE43+11Zn	10.90	1.80	0.73	0.28	0.12	balance
WE43+14Zn	13.80	3.06	1.04	0.29	0.10	balance
WE43+26Zn	25.82	3.02	1.16	0.27	0.17	balance

Table 2. Compositions of Mg-Zn-Al-based alloys in wt.%.

	Zn	Al	Ca	Mg
Mg5Zn3Al	5.3	3.2	0.1	balance
Mg12Zn3Al	11.9	3.1	-	balance

Digital positron lifetime spectrometer [4] with excellent time resolution of 145 ps was used for positron lifetime measurements. The resolution of the spectrometer was calculated as FWHM of resolution function obtained from fitting of positron lifetime spectrum of a well-annealed Mg reference sample. Each measured spectrum consisted of at least 10^7 annihilation events. The $^{22}Na_2CO_3$ positron source deposited on 2 μm thick mylar foil exhibits source contribution consisting of two components with lifetimes 368 ps and 1.5 ns and relative intensities 7% and 1%, respectively, which was always subtracted from the spectra.

The measurements of Vickers microhardness (HV) were performed using a STRUERS Duramin 300 hardness tester with a load of 100 g applied for 10 s.

In-situ measurement of synchrotron X-ray diffraction during heat treatment was used to monitor evolution of phase composition at elevated temperatures. Sample of as-cast WE43+14Zn alloy was subjected to linear heating from room temperature to 425°C. Subsequently the sample was cooled down back to room temperature. Heating and cooling rates were 5 K/min, wavelength of X-ray radiation was set to $\lambda = 0.142$ Å.

3. Results

Results of positron lifetime measurements of as-cast alloys are shown in table 3. The WE43 alloy, which does not contain quasicrystalline I-phase, exhibits single component spectrum with lifetime $\tau_1 = (223.9 \pm 0.3)$ ps which agrees well with the bulk positron lifetime in well-annealed Mg [5]. All other alloys, which contain quasicrystalline phase, exhibit two component spectra, see table 3. The first component comes from annihilation of free positrons while the second component can be

Table 3. Results of positron lifetime measurements of as cast samples.

	State	τ_1 (ps)	I_1 (%)	τ_2 (ps)	I_2 (%)
WE43	as-cast	223.9(3)	100	-	-
WE43+26Zn	as-cast	192(4)	56(3)	302(4)	44(3)
WE43+14Zn	as-cast	187(2)	58(2)	302(3)	42(2)
WE43+11Zn	as-cast	201(2)	67(3)	296(5)	33(3)
Mg5Zn3Al	as-cast	219(1)	96(1)	290(10)	4(1)
Mg12Zn3Al	as-cast	217.3(5)	93.9(5)	300(10)	6.1(5)

attributed to annihilation of positrons trapped at defects. The lifetime of the second component $\tau_2 \approx$ 300 ps provides characteristic signature of misfit defects associated with quasicrystalline phase [3].

Table 4. Results of positron lifetime measurements of WE43+11Zn samples annealed at 500°C.

	State	τ_1 (ps)	I_1 (%)	τ_2 (ps)	I_2 (%)
WE43+11Zn	as-cast	201(2)	67(3)	296(5)	33(3)
WE43+11Zn	500°C/1 h Q	219(1)	96(1)	290(10)	4(1)
WE43+11Zn	500°C/1 h SC	217.3(5)	93.9(5)	300(10)	6.1(5)

The effect of different cooling rates after isothermal annealing was studied on WE43+11Zn alloy. Two samples were annealed at 500°C for 1 hour, one was quenched into the water (Q) while the second one was slowly cooled down with furnace (SC). Results of positron lifetime measurements presented in table 4 revealed that misfit defects remain present in the alloys after annealing at 500°C. However, the fraction of positrons trapped at defects significantly decreases after annealing, especially in the slowly cooled sample.

Table 5. Results of positron lifetime measurements of isochronally annealed WE43+11Zn alloy.

	State	τ_1 (ps)	I_1 (%)	τ_2 (ps)	I_2 (%)
WE43+11Zn	as-cast	201(2)	67(3)	296(5)	33(3)
WE43+11Zn	160°C	180(10)	48(4)	278(9)	52(4)
WE43+11Zn	300°C	203(9)	70(5)	298(7)	30(5)
WE43+11Zn	440°C	210(10)	75(5)	300(10)	25(5)

The Mg-Zn-Y-based alloys were isochronally annealed with step 20 K/20 min to monitor the development of physical properties with increasing annealing temperature. The evolution of microhardness during isochronal annealing is shown in figure 1. Hardness has generally decreasing trend but there are two notable hardening peaks at ~ 160°C and ~ 440°C.

Based on the development of hardness, annealing temperatures of 160°C, 300°C and 440°C (indicated in figure 1 by arrows) were selected for positron lifetime spectroscopy measurements.

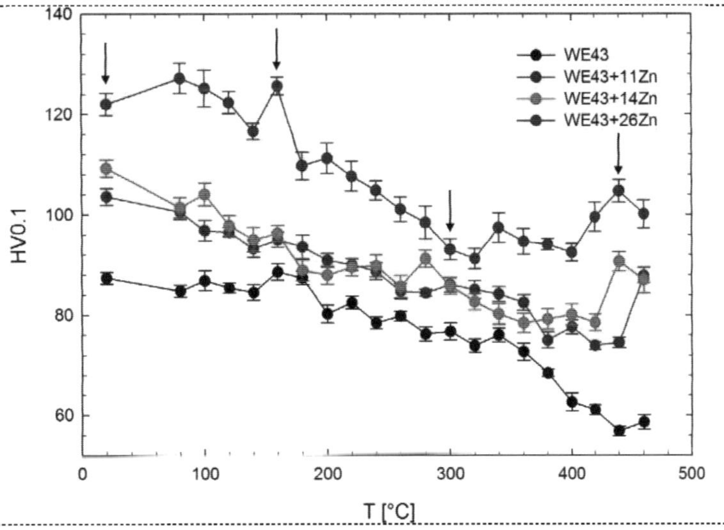

Figure 1. Evolution of microhardness during isochronal annealing.

Journal of Physics: Conference Series **674** (2016) 012005 doi:10.1088/1742-6596/674/1/012005

The results of these measurements for WE43+11Zn alloy are summarized in table 5. After annealing at 160°C, the intensity of the second component increased and its lifetime decreased by roughly 20 ps. It has to be noted that positron lifetime of ~ 260 ps was observed in binary Mg-Zn alloy after annealing at 220°C [6]. This lifetime was attributed to positrons trapped at open-volume misfit defects at the interfaces between precipitates and Mg matrix. Therefore, most likely explanation of shortened lifetime τ_2 in WE43+11Zn alloy is that precipitation occurred after annealing at 160°C and components with lifetimes ~ 260 ps (misfit defect at precipitate-matrix interfaces) and ~ 300 ps (misfit defects associated with I-phase) cannot be separated due to limited experimental resolution. Instead, one component with a lifetime of $\tau_2 = (278 \pm 9)$ ps was observed. Precipitates formed at 160°C dissolve at higher annealing temperatures and the lifetime τ_2 of the second component again has value of ~ 300 ps. However, intensity of the second component is slightly decreased compared to the as-cast state.

Results of *in-situ* XRD studies of WE+14Zn alloy are presented in figure 2 as a 2D image in which individual XRD patterns for each temperature are shown as horizontal lines and intensity is converted to color. The vertical axis shows evolution of XRD patterns during heat treatment. In the as-cast sample, peaks corresponding to Mg matrix, icosahedral I-phase (Mg_3Zn_6Y), cubic W-phase ($Mg_3Zn_3Y_2$) and graphite crucible in which the sample was placed were identified. Peaks corresponding to the I-phase disappeared when temperature exceeded ~ 367°C, indicated by the red horizontal arrow in the figure. Diffuse bump in the background, testifying presence of liquid phase, also appeared at this temperature. Peaks of W-phase remained visible up to ~ 416°C, indicated by the blue horizontal arrow in the figure. During the cooling, peaks of W-phase and I-phase appeared at temperature ~ 416°C and ~ 367°C, respectively.

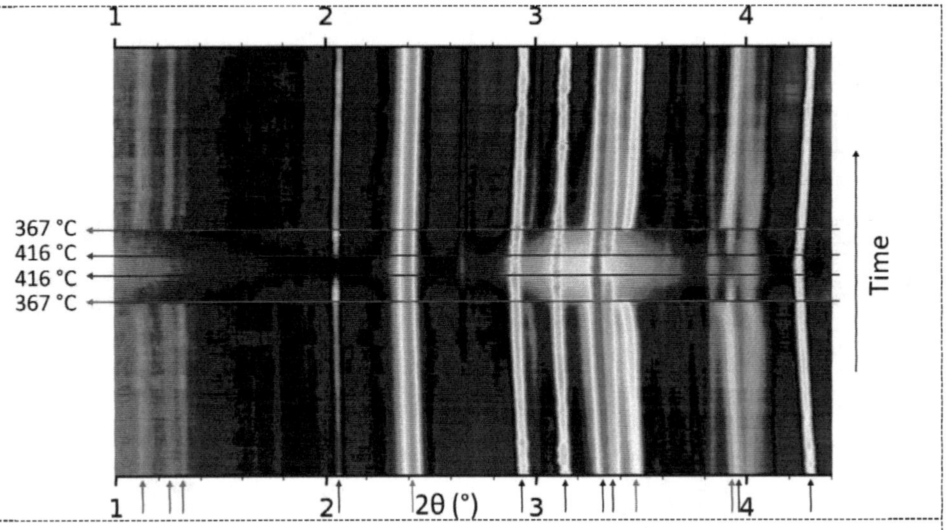

Figure 2. Results of *in situ* XRD measurement during linear heat up and cool down from room temperature to 425°C. Each horizontal line represents diffraction pattern at some temperature. Intensity was converted to the color scale and increases from black to red. Most pronounced peaks of Mg matrix (black), I-phase (red), W-phase (blue) and graphite crucible (gray) are marked by colored arrows.

Because of lattice expansion the diffraction peaks shift to lower diffraction angles with increasing temperature their temperature evolution during heating and subsequent cooling in figure 2 forms typical "V" shape. The diffraction peaks corresponding to I-phase show different behavior than other peaks. Contrary to other phases the positions of the peaks corresponding to I-phase do not change linearly with temperature but exhibits nonlinear dependence which can be seen as slightly bent curves in figure 2. The relative lattice expansion with respect to room temperature was calculated from fitted positions of the diffraction peaks and is plotted in figure 3. Thermal expansion of Mg matrix and graphite is approximately linear with temperature. On the other hand, thermal expansion of I-phase exhibits a strong nonlinearity indicating that the coefficient of thermal expansion of I-phase remarkably changes around 310°C.

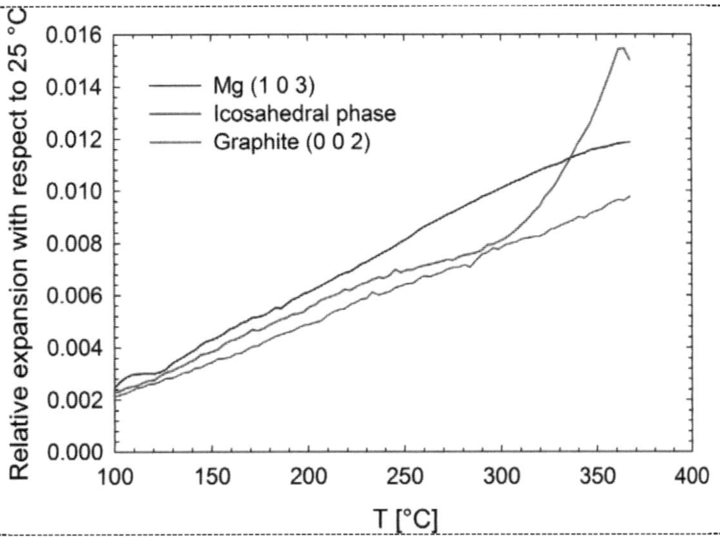

Figure 3. Relative thermal expansion with respect to 25°C calculated from XRD data.

4. Discussion

Results of positron lifetime measurements show that all studied alloys containing quasicrystalline phase exhibit two component spectrum with lifetime of the second component $\tau_2 \approx 300$ ps. This lifetime therefore characterizes misfit defects which are uniquely associated with quasicrystalline phase in Mg alloys. These defects are most likely situated at the interface of quasicrystalline phase and Mg matrix. Due to incommensurability of their lattices, misfit defects must be created at their interface. In order to partially retain the coherency between I-phase and α-Mg matrix steps or ledges are introduced periodically along the interface and misfit defects are located at these steps along the interface. Presence of these ledges was confirmed by high resolution transmission electron microscopy [7].

Size of the misfit defects can be estimated by comparison of the $\tau_2 \approx 300$ ps with the positron lifetime for a Mg monovacancy. Theoretically calculated lifetime for a Mg monovacancy is ≈ 300 ps [5]. This would suggest that misfit defects associated with I-phase are approximately of the same size as Mg monovacancy. On the other hand, the experimental positron lifetime reported for Mg monovacancy is ≈ 255 ps [8]. Hence, according to these experimental results, the misfit defects associated with I-phase are larger than Mg monovacancy. The discrepancy between experimental results and theoretical calculations of positron lifetime for Mg monovacancy is further discussed in reference [9].

Although annealing at 500°C increased the volume fraction of I-phase in grain boundary phase [10], the intensity of the second component in positron lifetime spectrum decreased after annealing. This is due to the fact, that finely dispersed precipitates of I-phase are present in the in Mg matrix of the as-cast alloy and most of these small precipitates are dissolved during annealing. Hence, the content of I-phase in grain boundary phase is enhanced by annealing at 500°C but its concentration in Mg grains is decreased due to dissolution of the fine I-phase precipitates. This leads to a reduction of the fraction of positrons trapped at misfit defects associated with I-phase. Decrease in the intensity of the second component in positron lifetime spectra after isochronal annealing at temperatures 300°C and 440°C is caused by the same reason.

In situ XRD investigations were performed to study phase transformations in WE43+14Zn alloy during linear heating. The temperatures of formation, i.e. ~ 367°C for the I-phase and ~ 416°C for the W-phase, determined by XRD differ significantly from the values obtained by differential thermal analysis, which are ~ 450°C [7] and ~ 510°C [11]. Our differential scanning calorimetry measurements of WE+14Zn alloy revealed presence of four endothermic peaks at temperatures 330-360°C, ~ 420°C, ~ 445°C and ~ 505°C. First two peaks could be attributed to melting of I-phase and W-phase respectively. The most pronounced peak located at 445°C might be due to melting of whole grain boundary region. However, using this interpretation, the origin of the last peak located around 505°C remains unknown. Further investigations are clearly needed to definitively assign phase transitions observed by X-ray diffraction to peaks measured by differential scanning calorimetry.

Thermal lattice expansion of I-phase shows significant nonlinearity with increase of the thermal expansion around 310°C. This is most likely linked to some structural changes in I-phase occurring in this temperature range. Additional work is needed to elucidate the cause of this effect.

5. Conclusions

Misfit defects associated with interfaces between I-phase and Mg matrix were observed in Mg-Zn-Y and Mg-Zn-Al alloys. Density of defects decreases after annealing at 500°C due to decrease of density of precipitates in Mg matrix and grain growth.

Nonlinear thermal expansion of I-phase lattice was observed above 310°C, which indicates some structural change in the I-phase. Temperatures of phase transformations occurring in WE43+14Zn alloy were determined by in situ XRD during linear heating.

6. Acknowledgements

This work was supported by the Czech Science Foundation (project P108/10/0648), the grant agency of Charles University (project 566012) and by the grant SVV-2014-260091.

References

[1] Luo Z, Zhang S, Tang Y and Zhao D 1993 *Scripta Metall. Mater.* **28** 1513
[2] Vekilov Y K and Chernikov M A 2010 *Physics - Uspekhi* **53** 537
[3] Čížek J, Vlček M, Smola B, Stulíková I, Procházka I, Kužel R, Jäger A and Lejček P 2012 *Scripta Mater.* **66** 630
[4] Bečvář F, Čížek J, Procházka I and Janotová J 2005 *Nucl. Instr. Meth. Phys. Res. A* **539** 372
[5] Robles J M C, Ogando E and Plazaola F 2007 *J. Phys.: Condens. Matter* **19** 176222
[6] Hruška P, Čížek J, Vlček M, Melikhova O and Procházka I 2014 *Acta Phys. Pol. A* **125** 718
[7] Bae D, Kim S, Kim D and Kim W 2002 *Acta Mater.* **50** 2343
[8] Hautojärvi P, Johansson J, Vehanen A and Yli-Kauppila J 1982 *Appl. Phys. A* **27** 49
[9] Vlček M, Čížek J, Melikhova O, Hruška P, Procházka I, Vlach M, Stulíková I and Smola B *J. Phys. Conf. Ser.* Submitted
[10] Melikhova O, Čížek J, Vlček M, Lukáč F, Vlach M, Smola B, Stulíková I, Kužel R, Procházka I, Kudrnová H and Jäger A 2014 *Acta Phys. Pol. A* **125** 718
[11] Zhang Y, Zeng X, Liu L, Lu C, Zhou H, Li Q and Zhu Y 2004 *Mater. Sci. Eng. A* **373** 320

Journal of Physics: Conference Series **674** (2016) 012006 doi:10.1088/1742-6596/674/1/012006

Vacancy clustering behavior in hydrogen-charged martensitic steel AISI 410 under tensile deformation

K Sugita, Y Mutou, Y Shirai

Department of Materials Science and Engineering, Graduate School of Engineering, Kyoto University, Yoshida, Sakyo-ku, Kyoto 606-8501, Japan

E-mail: sugita.kazuki.3v@kyoto-u.ac.jp

Abstract. The formation and accumulation of defects under tensile deformation of hydrogen-charged AISI 410 martensitic steels were investigated by using positron lifetime spectroscopy. During the deformation process, dislocations and vacancy-clusters were introduced and increased with increasing strains. Between hydrogen-charged and uncharged samples with the same tensile strains there was no significant difference in the dislocation density and mono-vacancy equivalent vacancy density.

1. Introduction

Hydrogen embrittlement is a serious problem responsible primarily for delayed failures that frequently occurs when high-strength martensitic steels and other metals absorb hydrogen. It is characterized by mechanical properties degradation due to hydrogen ingress, such as ductility reduction. These behaviors imply that the hydrogen embrittlement is directly correlated with lattice defects and most models on hydrogen embrittlement are based on hydrogen-defect interactions. However, the interaction between hydrogen and defects during deformation has not been revealed enough yet. Positron lifetime spectroscopy is specifically sensitive to open-volume defects, such as dislocations, vacancies and vacancy clusters. In particular, it has a distinctive advantage in detecting vacancy or vacancy cluster, and evaluating in vacancy densities through multi-component spectrum analysis based on the multi-state positron trapping model.

AISI 410 steel is one of the most popular 400 series martensitic stainless steels, and is widely used for the high strength and corrosion resistant applications. In this study, the formation and accumulation of defects in AISI 410 steel under tensile deformation were investigated by using positron lifetime spectroscopy.

2. Experimental method

The chemical composition of the commercial AISI 410 steel plates used in this study is provided in Table 1. The composition is within the typical composition range of AISI 410 steel. The steel plates with a thickness of 5 mm were machined to the tensile test specimen geometry with a gauge width of 5 mm, a gauge length of 30 mm and a thickness of 5 mm. Then tensile test specimens were polished up to grit #2000 with waterproof abrasive papers. The hydrogen charging was conducted by soaking for

Table 1. Chemical composition (wt%) of steel plates. The composition range of AISI 410 steel is also shown for comparison.

	Fe	Cr	Mn	Si	C	Ni	P	S
Steel studied	Balance	12.73	0.45	0.38	0.14	0.07	0.022	0.006
AISI 410	Balance	11.5-13.5	1 max	1 max	0.15 max	0.75 max	0.04 max	0.03 max

48h at 50°C in a 20wt% ammonium thiocyanate (NH4SCN) solution. The diffusible hydrogen content in the hydrogen-charged samples was evaluated as 2.2 wtppm based on thermal desorption analysis.

Tensile tests were carried out at room temperature using a servo-hydraulic universal testing machine (EHF-EV101k2-A20-1A, Shimadzu Corporation) at a strain rate of 8.0×10^{-6} s^{-1} for hydrogen-charged and uncharged samples, and fractured and interrupted samples were prepared. In order to prevent hydrogen release from hydrogen-charged samples during the tensile deformation, the tensile test specimens were soaked in the NH_4SCN solution. A pair of samples for positron lifetime measurements was obtained from the gauge part of the tensile test specimens. The samples obtained were mechanically polished and then electrolytically polished by using an ammonium chloride solution. Positron lifetime measurements were carried out by using a digital oscilloscope system with photomultiplier tubes mounted with BaF_2 scinitillators, having a time resolution (FWHM) of 175 ps. The positron lifetime measurements require data acquisition for approximately 9 hours with a Na-22 positron source of 0.7MBq activity in order to acquire 3 million counts in the positron lifetime spectrum. The measured spectra were analyzed using the programs RESOLUTION [1] and POSITRONFIT Extended [2].

3. Results

The engineering stress-strain curves for hydrogen-free and hydrogen-charged samples are shown in Figure 1. The tensile elongation was obviously reduced by hydrogen charging. The positron lifetimes τ_{one} obtained by one-component fitting of lifetime spectra are shown in Figure 2. The positron lifetime for hydrogen-charged samples showed higher values compared to uncharged samples.

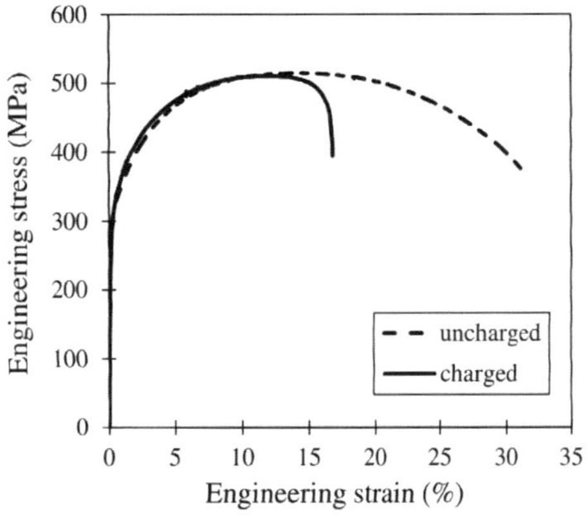

Figure 1. Engineering stress-strain curves for the hydrogen-charged and uncharged samples.

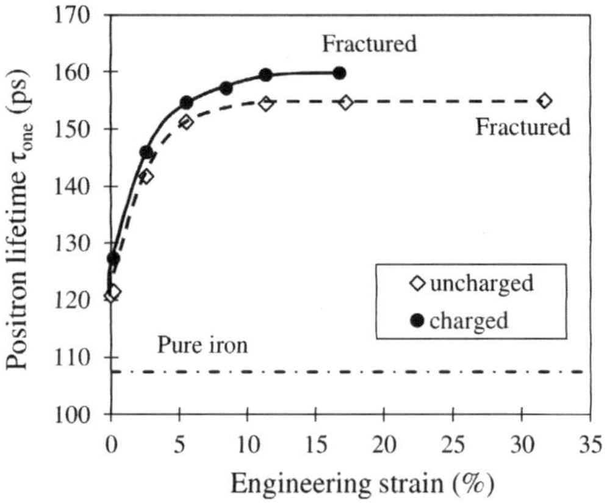

Figure 2. Positron lifetime changes with tensile strain for the hydrogen-charged and uncharged samples.

Journal of Physics: Conference Series **674** (2016) 012006 doi:10.1088/1742-6596/674/1/012006

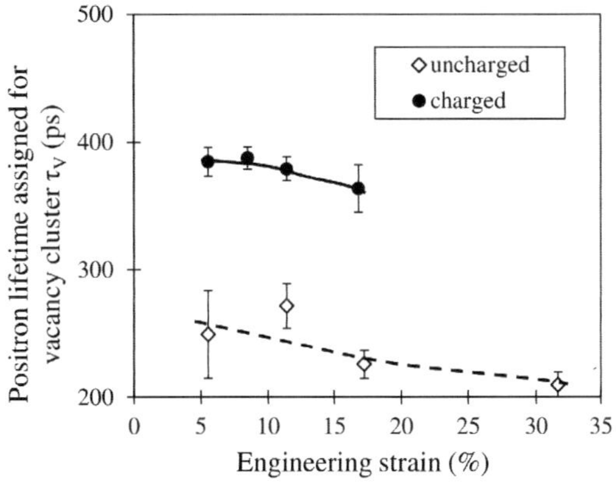

Figure 3. Changes of posiron lifetime assigned for vacancy-cluster with engineering strains for hydrogen-charged and uncharged samples.

The positron lifetimes for hydrogen-charged and uncharged samples tended to increase with increasing tensile strains, and at the final fracture stages they exceeded 150 ps, which is corresponding to the positron annihilation lifetime [3, 4] at the dislocation. This implies that the vacancy-type defect which has larger positron lifetime than the dislocation was introduced during the deformation process. Then the measured positron lifetime spectra were decomposed to the three components corresponding to the positron annihilation in *free state*, at *dislocations* and the other *vacancy-type* defects which have longer positron lifetime than dislocations using the positron trapping model [5].

The positron lifetimes assigned for *dislocations* and the other *vacancy-type* defects were 145-150 ps and 200-400 ps, respectively. The positron lifetime of 145-150 ps was within the range of previously reported values for dislocations in pure iron and iron-based alloys (150 ps [3, 4], 142 ps (screw[6, 7]), 165 ps (edge[6, 7]). The longer positron lifetime of 200-400 ps is related to the calculated value for the vacancy cluster consisting of 2-15 vacancies in pure iron[8]. So, the shorter defect components were assigned as for dislocations and the longer as for vacancy clusters. Figure 3 shows the changes of posiron lifetime assigned for vacancy-clusters with engineering strains for the hydrogen-charged and uncharged samples. The positron lifetime assigned for vacancy-cluster was 360-390 ps for the hydrogen-charged samples, which corresponds to the vacancy clusters consisting of 15 vacancies as reported by the theoretical study on pure iron [8]. On the other hand, in the hydrogen-free samples the number of vacancies in vacancy-clusters stayed at 2-4. These results definitely show that the vacancy cluster size formed during the tensile deformation is significantly affected by hydrogen charging.

Then the defect densities were evaluated from the three component analysis data by using the following equations based on the trapping model;

$$C_d = \frac{1/\tau_f - 1/\tau_d + I_v(1/\tau_d - 1/\tau_v)}{1 - I_d - I_v} \times \frac{I_d}{\mu_d} \qquad (1)$$

$$C_v = \frac{1/\tau_f - 1/\tau_v + I_d(1/\tau_v - 1/\tau_d)}{1 - I_v - I_d} \times \frac{I_v}{\mu_v} \qquad (2)$$

where C is a defect density, I is a relative intensity of the defect component, τ is positron lifetime, μ is the specific trapping rate at the defect. The index f stands for *free-state*, d for *dislocations* and v for *vacancy clusters*. When the defect density is very high and $1\text{-}I_v\text{-}I_d$ is very small, the dislocation density was determined by reference to the evaluation results of subsequently annealed samples up to 200°C as described later. The positron lifetime for dislocation τ_d was analytically-determined as 150 ps, based on the data obtained for the subsequently annealed samples, and fixed in the multicomponent analysis. The specific trapping rates used in the evaluation were as follows; μ_d=5.1×10^{-5} m^{2}s^{-1} (experimental value reported in the reference [7]) and $\mu_v \approx n \cdot \mu_{1v}$=1.1$n$×10^{15} s^{-1} for the vacancy cluster consisting of n vacancies (μ_{1v}=1.1×10^{15} s^{-1}, experimental value reported as for mono-vacancy in pure iron [9]). The proportional relationship of the positron trapping rate reported in the theoretical work by Nieminen and Laakkonen [10] was assumed in the calculation. Figure 4 shows changes of (a) dislocation and (b) vacancy-cluster densities with engineering strain for the hydrogen-charged and uncharged samples. The change of mono-vacancy equivalent vacancy densities is also shown in the figure. The defect densities show significant increases with increasing strain especially in the early stage of the tensile deformations of hydrogen-charged and uncharged samples. However, there were no significant differences in the dislocation density and the equivalent vacancy density between the hydrogen charged and uncharged samples.

In order to investigate the thermal stability of vacancy-cluster, the fractured samples were subjected to the subsequent isochronal annealing for 1 hour at 100, 200, 300 and 400 °C. The measured positron

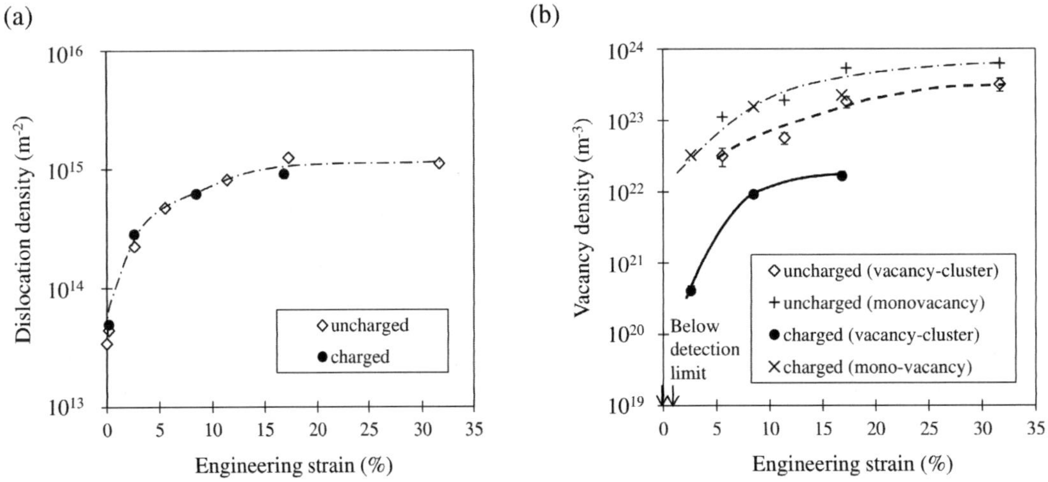

Figure 4. Changes of (a) dislocation and (b) vacancy-cluster densities with engineering strain for the hydrogen-charged and uncharged samples.

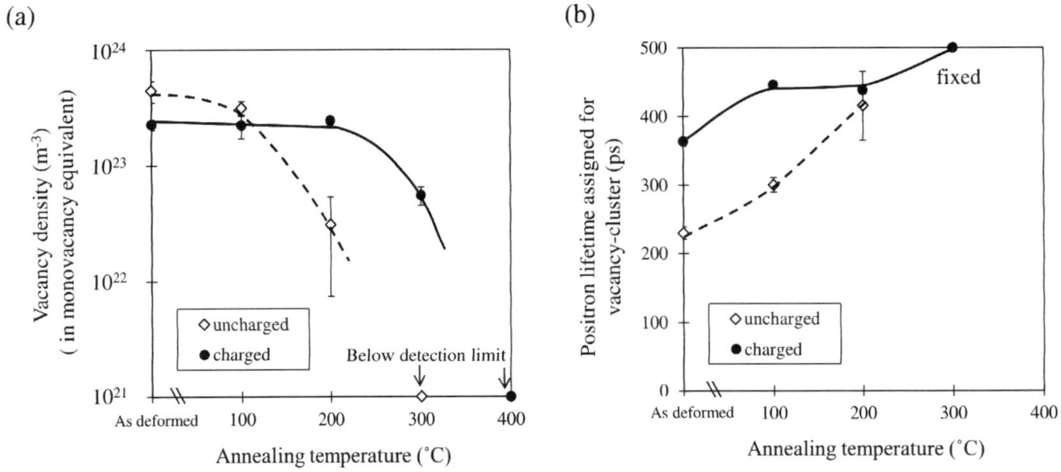

Figure 5. Changes of (a) Vacancy-cluster density and (b) positron lifetime with annealing temperature for the hydrogen-charged and uncharged samples.

lifetime spectra were decomposed to the three components and the defect densities were evaluated in the same manner as described above. In case of the hydrogen-charged sample annealed up to 300°C, the positron lifetime component assigned for vacancy cluster was about 600±100ps. So, in the multicomponent analysis, it was fixed to the value of 500 ps, which is corresponding to the saturation value of positrons lifetime for three-dimensional vacancy-cluster (voids) [11, 12]. Figure 5 shows the changes of (a) vacancy-cluster density and (b) positron lifetime assigned for vacancy-cluster with annealing temperature. It shows that the vacancy-cluster densities for the hydrogen-charged and uncharged samples decrease and vacancy-cluster sizes increase with increasing annealing temperature. This is typical vacancy clustering behavior where the excess vacancies introduced by irradiation or deformation migrate and partly merge into the larger vacancy clusters during annealing. The recovery stage of the vacancy-clusters shifts to higher temperature with hydrogen-charging, and this clearly shows that hydrogen has a significantly facilitative effect on the vacancy-clustering behavior during the deformation.

4. Discussion

In this work the facilitative effect on vacancy-clustering during tensile deformation due to hydrogen charging is obviously demonstrated and no large increase in the mono-vacancy equivalent vacancy density is also shown. These results show a partly consistent with Hydrogen Enhanced Strain Induced Vacancy Formation (HESIV) mechanism which claims that the formation of vacancies and nanovoids are facilitated under strain in the presence of hydrogen, and leads to ductile crack growth by deformation localization [13]. By thermal desorption studies [14], the increase of desorbed hydrogen contents are observed and the authors claim that the formation of vacancy-type defects are facilitated by hydrogen charging. Their results indicate that desorbed hydrogen contents obtained by the thermal desorption analysis are significantly affected not only by vacancy density but also by vacancy-cluster size. Thus the evaluation of vacancy-cluster behavior through the thermal desorption analysis has difficulty in distinguishing the density and cluster size effect. On the other hand, positron lifetime spectroscopy is the established method to detect mono-vacancy and/or small vacancy cluster, and to evaluate thier densities. Further positron annihilation studies are required to understand the effect of hydrogen on the vacancy behavior under tensile deformation.

5. Conclusion

In this study, the formation and accumulation of defects under tensile deformation of hydrogen-charged AISI 410 martensitic steels were investigated by using positron lifetime spectroscopy. During the deformation process, dislocations and vacancy-clusters were introduced and increased with increasing strains. Between hydrogen-charged and uncharged samples with the same tensile strains there was no significant difference in dislocation density and the mono-vacancy equivalent vacancy density.

References

[1] Kirkegaard P, Eldrup M, Mogensen O E and Pedersen N J 1981 *Comput. Phys. Commun.* **23** 307
[2] Kirkegaard P and Eldrup M 1972 *Comput. Phys. Commun.* **3** 240
[3] Hidalgo C, Gonzalez-Doncel G, Linderoth S and San Juan J 1992 *Phys. Rev.* B **45** 7017
[4] Mostafa K M, De Baerdemaeker J, Calvillo P R, Van Caenegem N, Houbaert Y and Segers D 2008 *Acta Phys. Pol.* A **113** 471
[5] Brandt W 1967 *Positron annihilation* (New York: Academic Press) p 155
[6] Park Y K and Waber J T 1985 *Mater. Lett.* **3** 181
[7] Park Y K, W'aber T, Meshii M, Snead Jr. C L and Park C G 1986 *Phys. Rev.* B **34** 823
[8] Puska M J and Nieminen R M 1983 *J. Phys.* F **13** 333
[9] Vehanen A, Hautojärvi P, Johansson J and Yli-Kauppila J 1982 *Phys. Rev.* B **25** 762

[10] Nieminen R M and Laakkonen J 1979 J. *Appl. Phys.* **20** 181
[11] Puska M and Nieminen R M 1983 *J. Phys.* F **13** 333
[12] Eldrup M and Singh B N 1997 *J. Nucl. Mater.* **251** 132
[13] Nagumo M 2004 *Mater. Sci. Tech.* **20** 940
[14] Nagumo M, Nakamura M and Takai K 2001 *Metall. Mater. Trans.* A **32** 339

Ultra fine grained Ti prepared by severe plastic deformation

F Lukáč[1], J Čížek[1], J Knapp[1], I Procházka[1], P Zháňal[1], R K Islamgaliev[2]

[1]Charles University in Prague, Faculty of Mathematics and Physics, V Holešovičkách 2, 18000 Praha 8, Czech Republic
[2]Institute of Physics of Advanced Materials, Ufa State Aviation Technical University, 12 K. Marx Street, 450000 Ufa, Russian Federation

E-mail: frantisek.lukac@mff.cuni.cz

Abstract. The positron annihilation spectroscopy was employed for characterisation of defects in pure Ti with ultra fine grained (UFG) structure. UFG Ti samples were prepared by two techniques based on severe plastic deformation (SPD): (i) high pressure torsion (HPT) and (ii) equal channel angular pressing (ECAP). Although HPT is the most efficient technique for grain refinement, the size of HPT-deformed specimens is limited. On the other hand, ECAP is less efficient in grain refinement but enables to produce larger samples more suitable for industrial applications. Characterisation of defects by positron annihilation spectroscopy was accompanied by hardness testing in order to monitor the development of mechanical properties of UFG Ti.

1. Introduction

Broad industrial applicability of titanium originates in its high strength, relatively low weight and high temperature stability. Moreover, by alloying titanium with other elements one can produce alloys with excellent mechanical properties, which are frequently used in biomedical applications. Since some alloying elements lack biocompatibility, there is a strong request for improved mechanical properties in pure titanium. Therefore, the ultra fine grained (UFG) metals attract interest of many researchers due to their superior mechanical properties consisting in a favourable combination of high strength and sufficient ductility [1, 2]. High pressure torsion (HPT) [1] is the most efficient technique for grain refinement by severe plastic deformation (SPD) methods and enables to achieve UFG structure with grain size around ~100 nm. However, size of HPT-deformed samples is limited to disk shape specimens with thickness up to ~1 mm and diameter up to ~20 mm. Moreover, strain distribution is not uniform across the HPT-deformed samples but increases from the centre of the sample towards the periphery [1, 2]. The Equal channel angular pressing (ECAP) [1] provide more massive samples in the form of rods which are more suitable for industrial applications. However, grain refinement by ECAP is lower compared to the HPT technique as a rule. SPD introduces a high density of defects, namely dislocations generated by plastic deformation and vacancies created by crossing dislocations. These deformation-induced defects play the key role in properties of UFG materials [3,4]. Hence, characterisation of defects in UFG samples prepared by SPD represents a very important task. Positron annihilation spectroscopy (PAS) has been frequently employed for defect studies of UFG metals but these studies covered mainly metals with cubic structure, see e.g. Refs. [3,4]. In particular there is a lack of PAS studies of UFG pure Ti despite that fact that Ti represents an important structural material. In the present work the positron lifetime (LT) spectroscopy was employed for characterisation of defects and their spatial distribution in UFG Ti samples prepared by HPT and ECAP.

2. Experimental details

A pre-rolled bar of titanium of commercial purity Grade IV (99.0% Ti) was deformed by ECAP method. No annealing of the Ti bar was performed prior to ECAP processing. The Ti bar was pressed through channels with an intersection angle of 105°. ECAP deformation was performed at temperature of 300°C following the route B_C, which was found to be the most effective in grain refinement [5]. The samples prepared by ECAP were rods having circle cross-section with the diameter of 12.5 mm.

High purity titanium (99.7%) was used for high pressure torsion (HPT) method. Similarly to the previous case, no annealing of Ti samples was performed prior to HPT processing. Two rotating anvils were pressed against each other by a pressure of 6 GPa. One of the anvils was slowly rotating and 5 revolutions were made at room temperature. The HPT-deformed sample is disc shaped with the diameter of 7 mm and the thickness of 0.5 mm. HPT samples were cleaned thoroughly before measurements.

The Vickers microhardness (HV) was measured by STRUERS DuraScan 50 Hardness Tester with automatised indentation. The load of 100 g was applied for 10 seconds.

A digital LT spectrometer [6] with an excellent time resolution of 145 ps (FWHM $_{22}$Na) was employed for the LT investigations. A $^{22}Na_2CO_3$ positron source with the diameter less than 1 mm and with the activity of 1.5 MBq was deposited on a 2 mm thick Mylar foil. At least 10^7 positron annihilation events were accumulated in the LT spectra which were decomposed using a maximum likelihood based procedure [7]. The source contribution consisted of two weak components which come from positrons annihilated in the source spot and the covering Mylar foil and exhibit lifetimes of ~368 ps and ~1.5 ns and corresponding intensities of ~6 % and ~1 %, respectively.

3. Results and discussion

The homogeneity of commercially pure titanium prepared by ECAP method was checked in a transversal cut (i.e. in the cross-section of the sample perpendicular to the pressing direction) after one and ten ECAP passes. Figure 1a shows a colour coded HV map after one ECAP pass. Hardness is not uniform across the sample cross-section. Enhanced HV values observed at the periphery (outer radius) are most probably due to friction between the inner channel surface and the surface of the pressed sample. This makes deformation at the periphery higher than in the inner region of the sample. Fig. 1b shows a HV map of the sample deformed by 10 ECAP passes. One can see in the figure that the hardness of the sample substantially increased and moreover the homogeneity of hardness across the sample was remarkably improved.

Two reference samples were measured by LT spectroscopy: (i) high purity titanium annealed at 1000°C for 2 hour and slowly cooled and (ii) the same sample conventionally cold rolled at room temperature. The annealed sample exhibits a single component spectrum with positron lifetime of 144.6±0.6 ps. This value is in a good agreement with the Ti bulk lifetime of 147 ps obtained from theoretical calculations [8]. The cold rolled sample exhibits defect component with lifetime of 172±1 ps, which can be attributed to the lifetime of a positron trapped at a dislocation in titanium.

Figure 2 shows the development of positron lifetimes and relative intensities of the components resolved in the LT spectra of Ti samples subjected to various numbers of ECAP passes. The LT spectra of Ti deformed by one and two ECAP passes are three component consisting of: (i) a short-lived component with the lifetime τ_1 significantly lower than the bulk Ti lifetime; (ii) a longer component with the lifetime $\tau_2 \approx 170$ ps; and (iii) a long-lived component with the lifetime $\tau_3 \approx 255$ ps. The short-lived component with the lifetime τ_1 obviously comes from free positrons while the components with the lifetimes τ_2 and τ_3 represent contributions of positrons trapped at defects. ECAP deformation by four and more passes led to disappearance of the free positron component and virtually all positrons are trapped at defects. The lifetime $\tau_2 = 170$ ps corresponds well to the dislocation lifetime measured in the cold rolled reference sample and the component with lifetime τ_2 can be therefore attributed to positrons trapped at dislocations introduced by SPD. The lifetime $\tau_3 = 255$ ps is higher than the lifetime of positrons trapped at monovacancies in Ti [8]. Hence, the appearance of the component with the lifetime τ_3 suggests that larger open volume defects, namely vacancy clusters,

Journal of Physics: Conference Series **674** (2016) 012007 doi:10.1088/1742-6596/674/1/012007

were created by ECAP deformation. The ab initio calculations of positron lifetimes for vacancy clusters of various sizes in titanium were performed in Ref. [8]. By comparison of the lifetime τ_3 with the theoretical calculations in Ref. [8] one can conclude that vacancy clusters in ECAP-deformed titanium consist on the average of two vacancies.

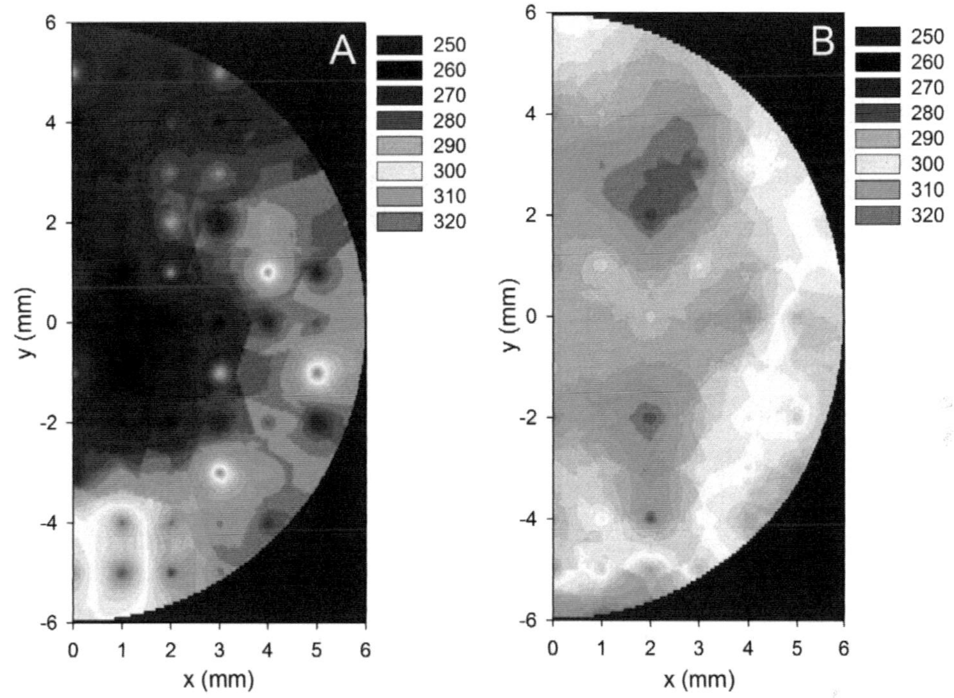

Figure 1. Colour coded maps of microhardness in the plane perpendicular to the direction of the ECAP pressing in commercially pure titanium (a) after one pass (b) after 10 passes. HV mapping was performed on a half of the sample cross-section.

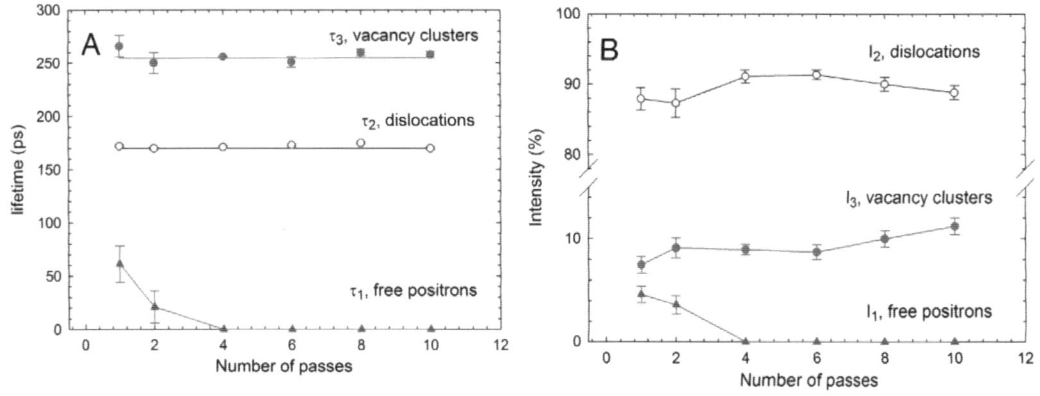

Figure 2. (a) The evolution of positron lifetime components distinguished in the spectra with respect to the number of ECAP presses. (b) The intensities of the positron lifetime components with respect to the number of ECAP presses.

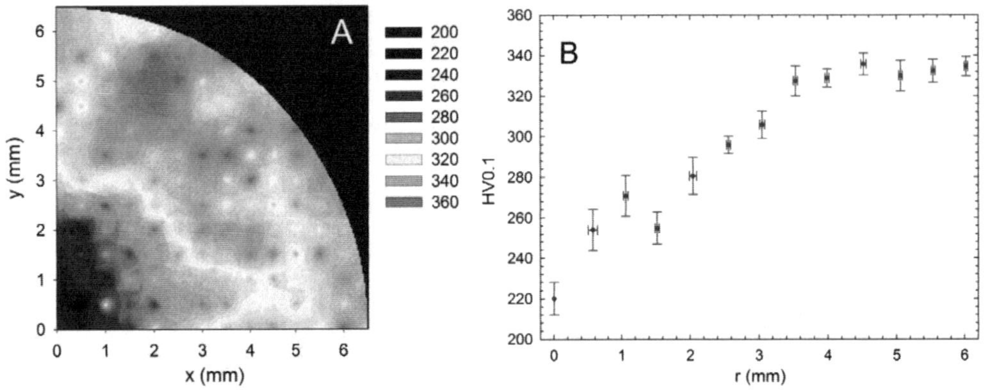

Figure 3. (a) A colour coded HV map of HPT-deformed titanium. Note that HV mapping was performed on a quarter of the sample disk. (b) The radial dependence of HV averaged over the whole angular range.

The homogeneity of high purity titanium prepared by HPT was examined by HV mapping and a colour coded HV map of the HPT-deformed sample is shown in Figure 3a. The dependence of HV values (averaged over the whole angular range) on the radial distance *r* from the centre of the specimen is plotted in Fig. 3b. HV increases from the centre of the sample towards the periphery. This is caused by non-uniform strain imposed by HPT deformation which linearly increases from the centre of the sample towards the periphery [1, 2].

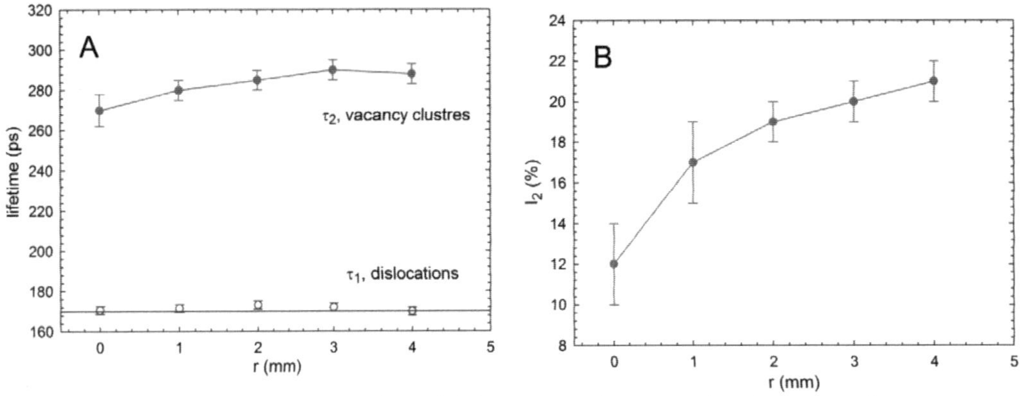

Figure 4. The development of (a) positron lifetimes and (b) the relative intensity of positrons trapped at vacancy clusters on the radial distance r from the centre of the HPT-deformed titanium.

LT spectroscopy was employed to examine the spatial distribution of defects across the HPT-deformed sample. The HPT-deformed titanium sample exhibits a two-component LT spectrum. The first component with lifetime $\tau_1 \approx 170$ ps comes from positrons trapped at dislocations while the second component with lifetime $\tau_2 \approx 280$ ps represents a contribution of positrons trapped at vacancy clusters. From comparison of the lifetime τ_2 with theoretical calculations in Ref. [8] one can deduce that vacancy clusters in HPT-deformed titanium consist on the average of three vacancies. Saturated positron trapping at dislocations and in vacancy clusters was observed in all points measured across the HPT-deformed sample. Fig. 4a shows the development of positron lifetimes τ_1, τ_2 with increasing radial distance from the centre of sample while the development of the relative intensity I_2 of positrons trapped at vacancy clusters is displayed in Fig. 4b. The intensity of the vacancy cluster component exhibits monotonic increase with the radial distance from the centre. It confirms that the concentration of vacancy clusters increases with r due to increasing strain. One can see in Fig. 4a that the lifetime τ_2 exhibits also a moderate increase with r. Hence, the average size of vacancy clusters slightly increases from the centre of the sample towards the periphery.

The lifetime of positrons trapped at vacancy clusters in HPT-deformed sample (~280 ps) is higher than the lifetime for vacancy clusters in the ECAP-deformed sample (~255 ps) indicating that open volume of vacancy clusters in the ECAP-deformed sample is lower. This is most probably caused by lower purity of the Ti sample deformed by ECAP. The commercial purity grade IV Ti used for ECAP deformation exhibits significantly higher concentration of interstitial oxygen compared to the high purity Ti used for HPT processing. The vacancy clusters introduced by ECAP processing are therefore likely coupled with interstitial oxygen atoms which reduce their open volume. The interaction of deformation-induced vacancies with interstitial oxygen is facilitated by the elevated temperature of ECAP processing (300 °C) which leads to a higher mobility of vacancies.

Comparing Figs. 2b and 4b one can notice that the intensity of positrons trapped at vacancy clusters is higher in the HPT-deformed sample than in the sample deformed by ECAP. It indicates that the HPT-deformed sample contains higher concentration of vacancy clusters than the sample deformed by ECAP. This is caused by the fact that that HPT imposes higher strain than ECAP processing. Note that no vacancy clusters could be detected in the Ti sample deformed by conventional cold rolling since the imposed strain was substantially lower than in case of SPD techniques and the concentration of deformation-induced vacancies is therefore too low.

4. Conclusions

UFG titanium samples produced by two different SPD methods (ECAP and HPT) were characterised by positron lifetime spectroscopy combined with microhardness mapping. Positron lifetime spectroscopy revealed that a high concentrations of dislocations are produced in ECAP and HPT-deformed samples of titanium. Moreover, vacancies produced during ECAP and HPT processing by crossing dislocations agglomerate into small vacancy clusters with different sizes. While the low temperature HPT processing produces vacancy clusters consisting on the average of three vacancies, the ECAP processing performed at elevated temperature induced vacancy clusters with smaller open volume most probably containing oxygen impurities. While the structure of ECAP-deformed sample becomes approximately uniform after sufficient number of passes, the HPT-deformed sample exhibits non-uniform structure due to non-uniform strain which increases from the centre towards the periphery. It is reflected by hardness increasing from the centre of the sample towards the periphery and the concentration and size of vacancy clusters which also increase with the radial distance from the centre.

5. Conclusions

This work was supported by the Czech Science Foundation (project P108/12/G043) and the Grant Agency of Charles University (Project No. 2002214).

References

[1] Zhilyaev A P and Langdon T G 2008 *Prog. Mater. Sci.* **53** 893
[2] Valiev R, Islamgaliev R and Alexandrov I 2000 *Progress in Materials Science* **45** 103
[3] Čížek J, Janeček M, Srba O, Kužel R, Barnovská Z, Procházka I and Dobatkin S 2011 *Acta Mater.* **59** 2322
[4] Su L, Lu C, He L, Zhang L, Guagliardo P, Tieu A, Samarin S, Williams J and Li H 2012 *Acta Mater.* **60** 4218
[5] Langdon T, Furukawa M, Nemoto M and Horita Z 2000 *J. Organomet. Chem.* **52** 30
[6] Bečvář F, Čížek J, Procházka I and Janotová J 2005 *Nucl. Instr. Meth. Phys. Res.* A **539** 372

[7] Procházka I, Novotný I and Be£vá° F 1997 *Mater. Sci. Forum* **255-257** 772
[8] Čížek J, Melikhova O, Barnovská Z, Procházka I and Islamgaliev R K 2013 *J. Phys.: Conf. Ser.* **443** 012008

Study of local structure in hyper-eutectic Zr-Cu-Al bulk glassy alloys by positron annihilation techniques

T Ishiyama[1], K Ishii[1], Y Yokoyama[2], T J Konno[2], A Iwase[1] and F Hori[1]

[1] Department of materials science, Osaka Prefecture University, 1-1 Gakuen-cho, Naka-ku, Sakai, Osaka 599-831, Japan
[2] Institute for Materials Research, Tohoku University, 2-1-1, Katahira, Aoba-ku, Sendai 980-8577, Japan

E-mail: horif@mtr.osakafu-u.ac.jp

Abstract. The Zr-Cu-Al bulk glassy (BG) alloy, which has amorphous structure, possesses various properties such as high strength and toughness with compositional dependence. In the present study, density, positron annihilation lifetime and coincidence Doppler Broadening measurement have been performed for various compositional hyper-eutectic Zr-Cu-Al BG alloys. The density of hyper-eutectic Zr-Cu-Al BG alloys increases with decreasing of Zr fraction. In contrast, positron lifetime for all compositional alloys is almost constant about 165 psec. In addition, the CDB ratio profile is almost the same for hyper-eutectic alloys. This unchanging trend of CDB ratio profile is quite different from that of hypo-eutectic BG alloys. These results reveal that different internal structure exists in hyper and hypo-eutectic BG alloys.

1. Introduction

Zr-Based bulk glassy alloy have attracted a great deal of attention over the past several years due to a high strength and a high toughness [1]. Recently, it was reported that the model for the glass forming ability is based on phase selection of the glass [2][3]. In addition, our co-workers have reported that the difference of the thermal volume coefficient and melting point the various compositions of Zr-based BG alloys [4]. Compositional dependence of these characteristics has been considered caused by the difference of the local structure such as clusters. Because those show tendencies of the change are different in the hypoeutectic and hypereutectic, it is considered that the local structure is different in the hypoeutectic and hypereutectic. Besides, the bulk glassy alloys have the free volume defined as frozen open volume and are not completely random structure, but forming clusters of short-range orderings and random areas [5]. So far, we have reported that there are a lot of Zr atoms around the free volume [6]. In the present study, it is reported that the evaluation based on microscopic structure measurement of the free volume for eutectic and hypoeutectic Zr-based bulk glassy alloy. However, the evaluation based on the free volume for hypereutectic have not been clarified yet. In this study, through the comparison with hypo-eutectic BG alloy, we studied the local structure around free volume in hyper-eutectic Zr-Cu-Al BG alloys by density measurement and positron annihilation techniques, such as positron annihilation lifetime and coincidence Doppler broadening (CDB) measurements. Positron lifetime is method for evaluating the electron density in free volume, and coincidence Doppler broadening measurement provides the information on the detailed electron distributions around free volume. In addition, in order to evaluate quantitatively, we calculated the S-parameter and W-parameter for the alloy composition from CDB spectra.

2. Experiments

The rod-shaped glassy alloys of hypo-eutectic $Zr_XCu_{90-X}Al_{10}$ (X=65, 60, 55), eutectic $Zr_{50}Cu_{40}Al_{10}$ and hypo-eutectic $Zr_XCu_{90-X}Al_{10}$ (X=49, 48, 47, 46, 45) were prepared by the tilt casting method in the arc furnace [7]. For positron annihilation measurements, these samples were cut into specimens with about 0.4 mm thickness. The as-prepared specimens were characterized by X-ray diffraction (XRD) measurements. In addition, we examined the glass transition temperature (T_g) and crystallization temperature (T_X) by differential scanning calorimetry (DSC), in order to confirm they are glass or not. The densities of these samples were also measured by Archimedes method at room temperature [8]. Positron lifetime and coincidence Doppler broadening (CDB) measurement have done for all specimens at room temperature. The positron annihilation lifetime spectra consist of 1.0×10^6 counts. All the positron annihilation lifetime spectra has analyzed by the RESOLUTION program. Each CDB spectra consists of more than 1.0×10^8 counts. The CDB ratio spectrums were obtained by normalizing the momentum distribution of each spectrum to that of Al metal. It is well established that the change in low electron momentum corresponds mainly to the free volume change, and the detailed profile at higher electron momentum region reflects the local structure and their atomic elements [9]. CDB spectra were characterized by S and W parameters. The S and W parameters were determined as the ratio of the γ-ray counts in the range of $|P_L|\leqq3.7\times10^{-3}$ m_0c and $1.1\times10^{-2}\leqq|P_L|\leqq1.7\times10^{-2}$ m_0c, respectively.

3. Results and discussion

Figure 1 shows the density for hyper-eutectic, hypo-eutectic and eutectic BG alloy. In this figure, the density linearly decreases with increasing Zr fraction. This systematic change with Zr fraction seems to be owing to total atomic density. That is, there is no different changing in macroscopic structure with Zr fraction. Figure 2 shows the results of positron annihilation lifetime measurement for all alloy composition BG alloy. The positron lifetime of eutectic and hypo-eutectic BG alloy agree with our previous reports [10, 11]. In this figure, positron lifetime remains stable for Zr fraction. Generally, it is well known that lifetime corresponds to the free volume size, such as vacancy [9]. It is considered that the free volume size in these alloys remains stable for Zr fraction.

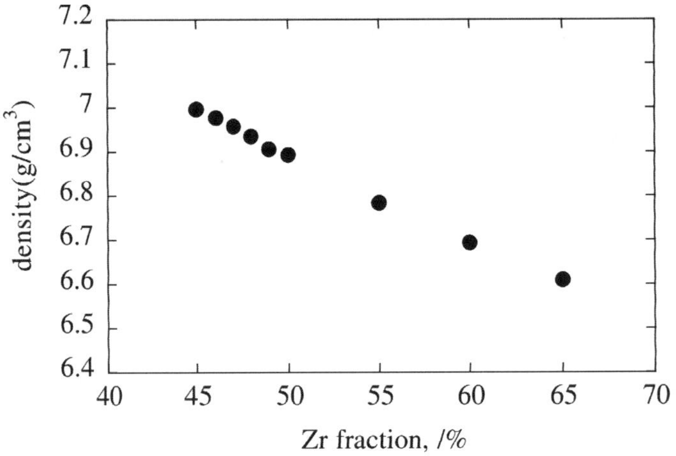

Figure 1. Density of Zr-Cu-Al bulk glassy alloys as a function of Zr fraction.

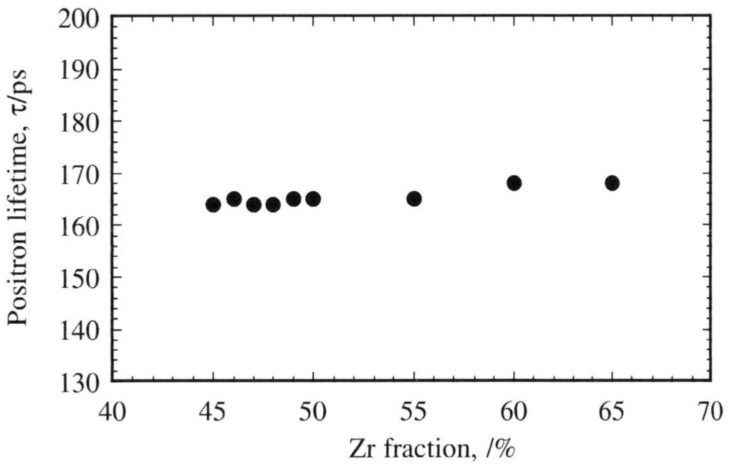

Figure 2. Positron annihilation lifetime of Zr-Cu-Al BG alloys as a function of Zr fraction.

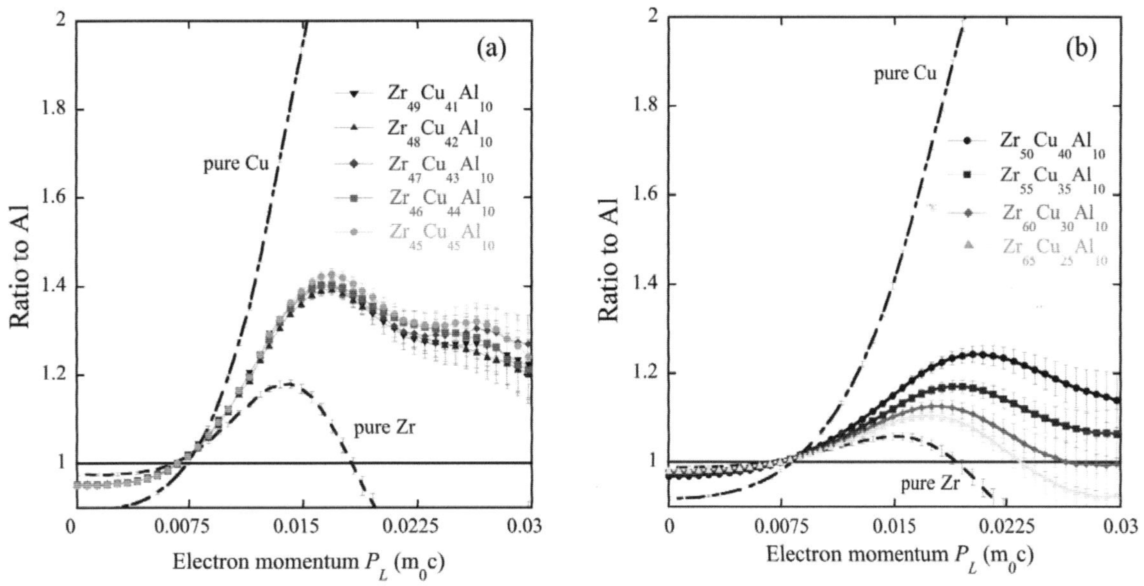

Figure 3. CDB ratio spectra of (a) hyper-eutectic and (b) hypo-eutectic bulk glassy alloy with those of Cu and Zr metals, expressed as a CDB ratio intensity of Al metal.

Figure 3 (a) and (b) shows the ratio spectra of CDB for all BG alloy with those of Cu and Zr metals, expressed as a ratio of CDB intensity of Al. The ratio profiles normalized by the CDB spectrum of pure Al are most clear in detail other than that of Fe and Zr. Also, only the broadening of ratio spectra for pure Zr in figure 3(a) and (b) is not consistent seemingly each other. This difference comes from the energy resolution difference by different measurement period, so that this difference has not influence to our discussions. In figure 3(a), the ratio spectra of hyper-eutectic alloys have little change with Zr fraction. In contrast, in case of the ratio spectra of hypo-eutectic alloys, systematically change is confirmed as shown in figure 3(b). With increasing Zr fraction, these ratio spectra approach to the ratio spectrum of pure Zr. This changing trend agrees well with our previous study [11]. It has reported that the icosahedral cluster is present around free volume in Zr-based BG alloys [11, 12]. Therefore, it is considered that the ratio of Zr atom around free volume increases with increasing Zr fraction. Both S-parameter and W-parameter as a function of Zr fraction for hyper-eutectic and hypo-eutectic BG alloys are shown in figure 4(a) and (b). In figure 3, the characteristic peak appears in high electron momentum around $0.015m_0c$, we selected this range as W-parameter. As seen in figure 4(a), S and W parameters for hyper-eutectic BG alloys are almost constant for Zr fraction, showing that the atomic element ratio around free volume does not change. In contrast, W-parameters in hypo-eutectic BG alloys change with Zr fraction as shown in figure 4(b). Comparing the results for hyper-eutectic and hypo-eutectic BG alloys, the changing trend of atomic element ratio is clearly different. This result can be considered as follows; that is the local ordered structure around free volume in hyper-eutectic BG alloys does not change even though Zr fraction increases. On the other hand, this local structure systematically changes with Zr fraction for hypo-eutectic BG alloys. This result is not consistent with the result of positron lifetime measurement. That is, the positron lifetime remains constant for Zr fraction, but W parameter changes only in case of hypo-eutectic BG alloys. This result can be supposed that the size of free volume becomes larger and surrounding Zr atom ratio increases around free volume with increasing of Zr fraction. Consequently, total electron density does not change at a free volume site in hypo-eutectic composition region.

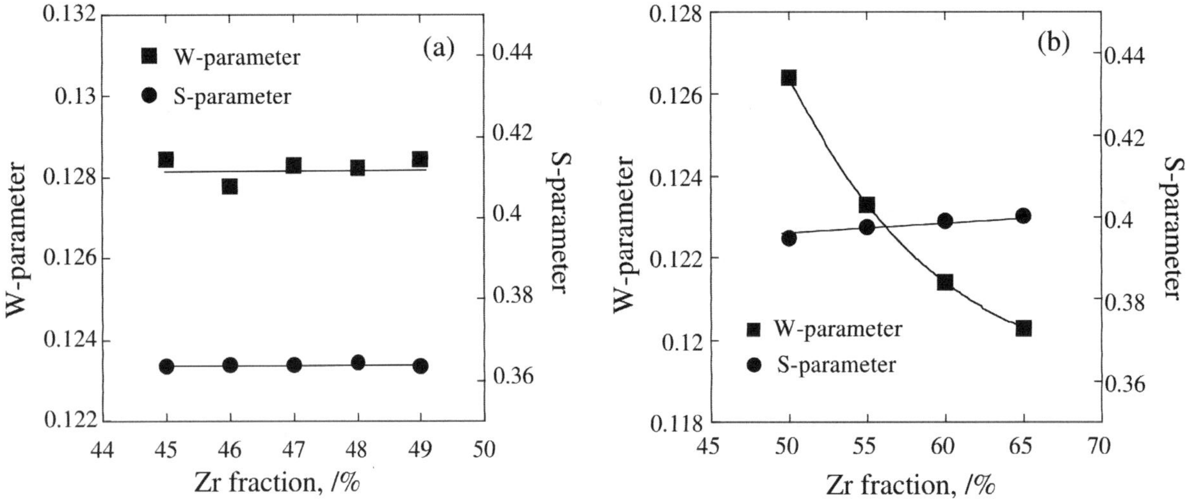

Figure 4. Each W-parameters and S-parameters of (a) hyper-eutectic and (b) hypo-eutectic bulk glassy alloy.

4. Summary

The difference of local structure for hyper-eutectic and hypo-eutectic Zr-Cu-Al bulk glassy alloys has been studied by positron annihilation lifetime and coincidence Doppler broadening measurements, and we obtained following conclusions. Regardless of the hyper-eutectic and hypo-eutectic, the density changes systematically with Zr fraction. On the other hand, the local structure around free volume in hyper-eutectic bulk glassy alloys has different trend from that in the hypo-eutectic one. The local structure around free volume for hyper-eutectic Zr-Cu-Al bulk glassy alloys is stable irrespective of hypo-eutectic Zr-Cu-Al bulk glassy alloys.

References

[1] Inoue A 2000 *Acta Materialla* **48** 279.
[2] Ma D, Tan H, Wang D and Li Y 2005 *Appl. Phys. Lett.* **86** 191906
[3] Kiminami C S, Sa Lisboal R D, de Oliveira M F, Bolfarini C and Botta W J 2007 *Mater. Trans.* **48** 1739
[4] Yokoyama Y, Ishikawa T, Okada J T, Watanabe Y, Nanao S and Inoue A 2009 *J. Non-Cryst. Solids* **355** 317
[5] Shimono M and Onodera H 2005 *Mater. Trans.* **46** 2830
[6] Ishii A, Hori F, Iwase A, Fukumoto Y, Yokoyama Y and Konno T J 2008 *Mater. Trans.* **49** 1975
[7] Yokoyama Y, Inoue K, Fukaura K 2002 *Mater. Trans.* **43** 2316
[8] Haruyama O, Nakayama Y, Wada R, Tokunaga H, Okada J, Ishikawa T and Yokoyama Y 2010 *Acta Mater.* **58** 1829
[9] Hautojarvi P 1979 *Topics in Current Physics* **12**
[10] Ishii A 2012 *doctorial thesis*
[11] Ishii A, Iwase A, Yokoyama Y, Konno T J, Kawasuso Y, Yabu-uchi A, Maekawa M, and Hori F, 2010 *J. Phys. Conf. Ser.* **225** 012020
[12] Kawamata T, Yokoyama Y, Saito M, Sugiyama K and Waseda Y 2010 *Mater. Trans.* **51** 1796

Journal of Physics: Conference Series **674** (2016) 012009 doi:10.1088/1742-6596/674/1/012009

Ni ion damage structures and hardness changes in austenitic stainless steels and their model alloys

K Sato[1], S. Ishioka[2], Q Xu[2], T Yoshiie[2], H Tsuchida[3], and A Itoh[3]

[1] Graduate School of Science and Engineering, Kagoshima University, Kagoshima-shi, Kagoshima 890-0065, Japan
[2] Research Reactor Institute, Kyoto University, Kumatori-cho, Osaka 590-0494, Japan
[3] Quantum Science and Engineering Center, Kyoto University, Uji, Kyoto 611-0011, Japan

E-mail: ksato@mech.kagoshima-u.ac.jp

Abstract. Defect structures of austenitic stainless steels and their model alloys after 6 MeV Ni ion irradiation were studied using positron annihilation lifetime measurements and Vickers hardness obtained from micro-hardness measurements. After irradiation at room temperature, a long lifetime of 150–180 ps was obtained and the Vickers hardness increased with increasing the irradiation dose. After irradiation at 573 K, all the alloys exhibited lifetime spectra that were not decomposed into two components. Their hardness also increased with the irradiation dose, but the enhancement was small compared with that at room temperature. Defects introduced by the ion irradiation could be detected near the surface using a Na-22 source. The hardness is correlated with the density and size of the defect clusters.

1. Introduction

To study the irradiation effects on nuclear materials, ions are commonly used as incident particles instead of neutrons (simulation irradiation). Ion irradiation provides many advantages such as high damage rates and easy setting of irradiation conditions, e.g., temperatures and stresses. One of the disadvantages is the localized formation of irradiation damage in the specimens because of a determined ion range. In the case of self ion irradiation of structural materials of nuclear systems such as steels with several MeV, the ion range is the order of microns and it is necessary to extract the bulk properties of the irradiation effects from the property changes at a thin area near the surface. Micro-hardness testing is one of the most widely used methods to measure mechanical properties of thin layers. This technique measures the penetration under different indentation loads to obtain an indentation curve. From these measured data, the mechanical properties such as hardness and elastic modulus can be evaluated [1].

Positron annihilation spectroscopy is a very powerful technique to detect open volume defects in materials because of their ability to trap positrons. Positron annihilation lifetime measurements and coincidence Doppler broadening measurements using Na-22 as a positron source have been adopted to study irradiation damage in bulk materials. Slow positron beam is the best way to detect irradiation defects in the near-surface region of ion-irradiated materials [2]. We have demonstrated, however, that it is possible to detect irradiation damages of 6 MeV Fe ion irradiated ferritic stainless steels by conventional sandwich configuration on positron source [3]. In the case of high ion irradiation, even if the damages are localized near the surface area, some of the positrons (the probability for the positrons to stop within ~3μm in depth in Fe is approximately 25%) are trapped there, thus the lifetime of the defects can be measured.

Austenitic stainless steels are important nuclear materials. Owing to the irradiation-induced swelling, they can experience, however, dimensional changes at very high dose neutron irradiation. It is known that the defect evolution strongly depends on the alloying elements [4-6]. In our previous studies, irradiation damages by electron irradiation and neutron irradiation were induced below 10^{-2} dpa and below 0.2 dpa, respectively, and the defect structural evolution before void swelling was studied [5-6]. In this study, defect structural evolutions of austenitic stainless steels and their model alloys were investigated at higher irradiation dose by Ni ion irradiation. Damaged structures and changes in the mechanical properties were investigated by micro-hardness tester and positron annihilation lifetime spectroscopy with Na-22.

2. Experimental

The test materials used in this study were Ni of 99.99% in purity, four model alloys, and four commercial alloys, as listed in Table 1. Ti-added modified SUS316SS (J) is an austenitic stainless steel with high Ni concentration to reduce void swelling, and designed for the first wall component of near term fusion devices [7-8]. The concentrations of the elements in the four model alloys, A, B, C, and D, were determined by the concentrations of the alloy constituents in J. For this experiment, thin disk specimens with the dimensions of 0.1 mm in thickness and 3 mm in diameter were machined and annealed at 1323 K and 10^{-4} Pa for 1 h, followed by cooling to room temperature. The cooling rate was sufficiently rapid, so that no precipitates were observed by transmission electron microscopy.

Specimens were irradiated by Ni ions (6 MeV) with an accelerator of Quantum Science and Engineering Center, Kyoto University at room temperature and at 573 K. The irradiation dose was 0.037, 0.37, 1.85, and 6.8 dpa (defect peak position). Figure 1 shows the implantation profiles of the Ni ions in Ni and the corresponding damage distributions calculated using the SRIM code [9]. The depth of the peak position of the defects is about 1.8 μm. We chose Ni as its crystal structure is the same and the lattice constant is almost the same as those of the austenitic stainless steels.

Table 1. Chemical compositions of the specimens.

Specimens	Composition
Ni	Pure Ni (99.99%)
A	Fe-16.1Cr-17.0Ni
B	Fe-15.4Cr-15.9Ni-2.68Mo-1.89Mn
C	Fe-15.3Cr-15.8Ni-2.66Mo-1.88Mn-0.53Si
D	Fe-15.3Cr-15.8Ni-2.66Mo-1.88Mn-0.53Si-0.24Ti
J	Ti added modified SUS316SS: Fe-15.3Cr-15.8Ni-2.66Mo-1.88Mn-0.53Si-0.24Ti-0.055C-0.024P
E	SUS316L SS: Fe-17.4Cr-12.05Ni-2.08Mo-0.85Mn-0.47Si-0.019C-0.027P-0.001S
F	SUS316SS: Fe-19.2Cr-12.58Ni-2.26Mo-1.82Mn-0.39Si-0.05C-0.019P-0.004S
G	SUS304SS: Fe-18.1Cr-9.08Ni-0.85Mn-0.48Si-0.05C-0.027P-0.0025S

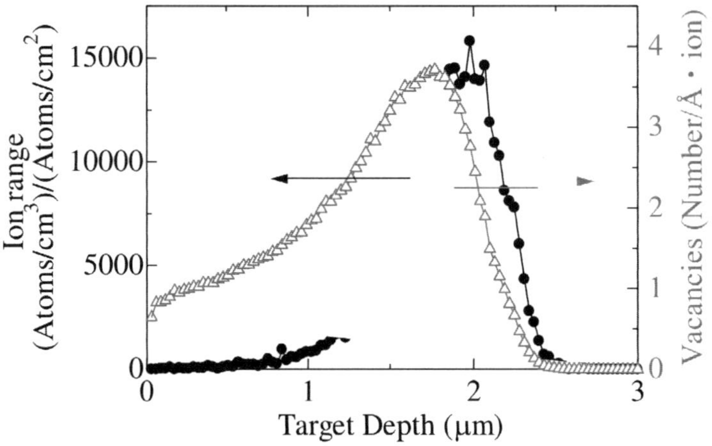

Figure 1. Ion range and damage profile of Ni implanted with 6 MeV Ni ions calculated by the SRIM code [9].

After irradiation, the damage structures were studied using positron annihilation lifetime measurements. The measurements were performed at room temperature using the fast-fast coincidence system, whose lifetime resolution was 190 ps (full width at half maximum, FWHM). The lifetime spectra were collected with a total count of about 10^6. The spectra were analyzed using the PALSfit program [10]. The total amount of residual defects can be derived from the mean positron lifetime. The amount of positron annihilation that occurs in the matrix can be determined from the short lifetime component. The size of the vacancy-type defects can be derived from the long lifetime component, and their density can be obtained from the relative intensity of the long lifetime component. The calculated positron annihilation lifetimes for single-vacancy and 13-vacancy clusters in Ni are 169 and 341 ps, respectively [11]. When the vacancy clusters grow, their positron lifetime becomes longer. A series of hardness tests was also performed by a micro-hardness tester (DUH201, Shimadzu Ltd.). The diamond indenter had a pyramidal shape with angles of 22° on each side with respect to the horizontal plane. The loading rate was 13.24 mN/s and the holding time was 5 s. The Vickers hardness was obtained by the following equation [12]:

$$H_v = 189.1\left(\frac{P}{d^2}\right), \tag{1}$$

where P is the load in mN (P = 10 mN) and d is the average length of the diagonal left by the indenter in μm.

3. Results and discussion

Figures 2–4 show the positron annihilation lifetime of austenitic stainless steels and their model alloys irradiated with Ni ions at room temperature and 0.037, 0.37, and 1.85 dpa, respectively. Before Ni ion irradiation, the lifetimes of Ni and the alloys were between 106 and 108 ps. A longer lifetime of approximately 150–200 ps was obtained after irradiation. The formation of large vacancy clusters was not detected. These tendencies closely resemble the results of neutron and electron irradiation experiments [5,6]. This is due to the fact that vacancies in these alloys are not mobile at room temperature (vacancy migration energies in Ni and austenitic stainless steels are 1.25 and 1.1–1.4 eV, respectively [13,14]). The lifetimes of commercial stainless steels were a little bit shorter than their model alloys. Figures 5–7 show the Vickers hardness. In all the specimens, the Vickers hardness increased with increasing the irradiation dose, owing to the fact that the amount of defects increases with the irradiation dose.

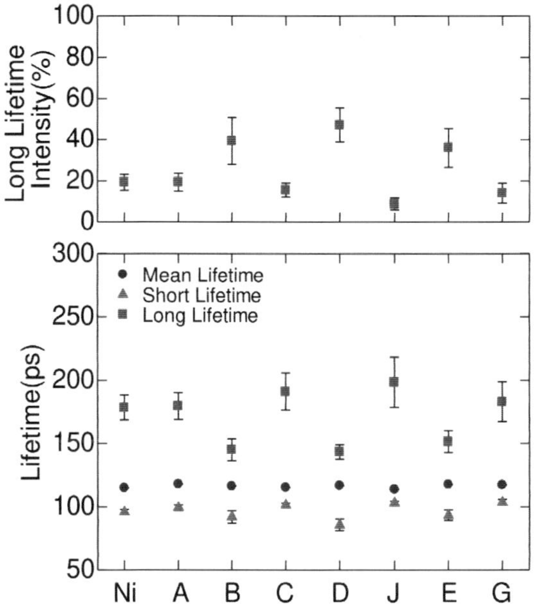

Figure 2. Positron annihilation lifetime of specimens irradiated with Ni ions at room temperature and 0.037 dpa.

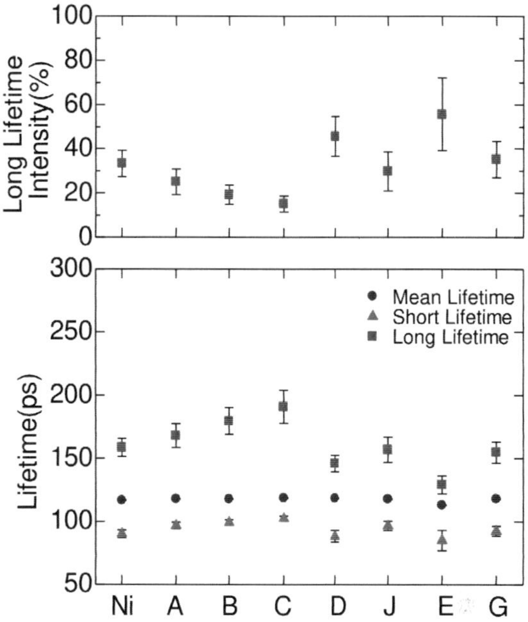

Figure 3. Positron annihilation lifetime of specimens irradiated with Ni ions at room temperature and 0.37 dpa.

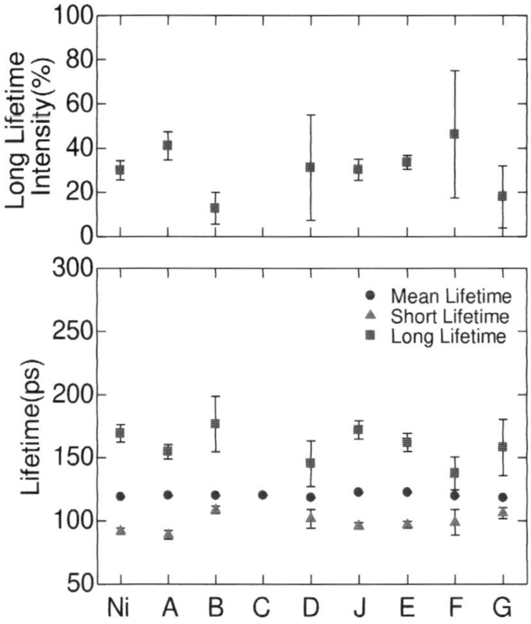

Figure 4. Positron annihilation lifetime of specimens irradiated with Ni ions at room temperature and 1.85 dpa.

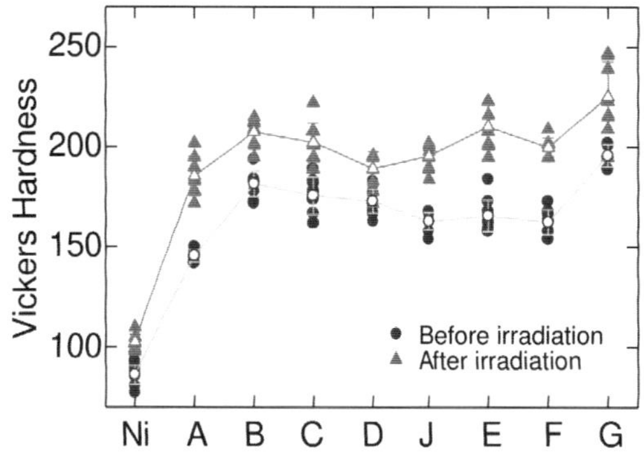

Figure 5. Vickers hardness of specimens irradiated with Ni ions at room temperature and 0.037 dpa.

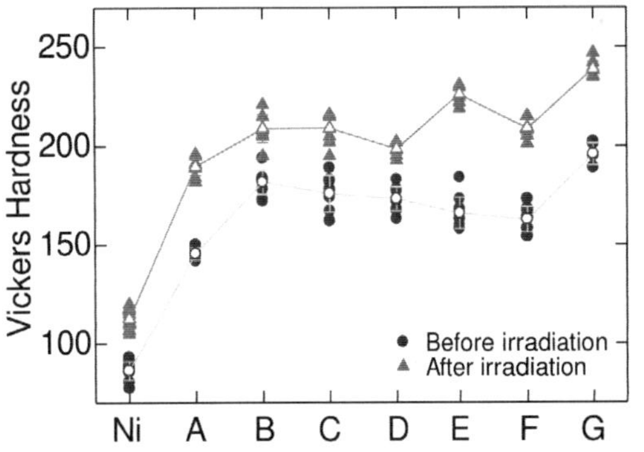

Figure 6. Vickers hardness of specimens irradiated with Ni ions at room temperature and 0.37 dpa.

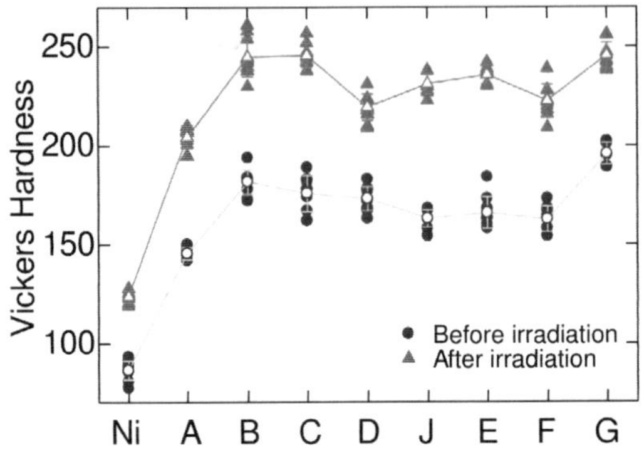

Figure 7. Vickers hardness of specimens irradiated with Ni ions at room temperature and 1.85 dpa.

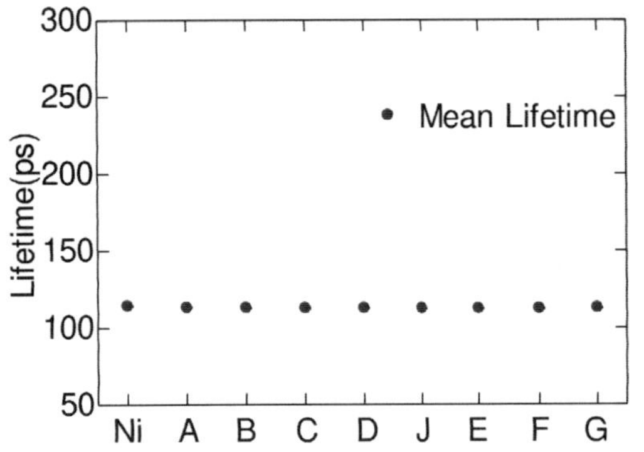

Figure 8. Positron annihilation lifetime of specimens irradiated with Ni ions at 573 K and 0.037 dpa.

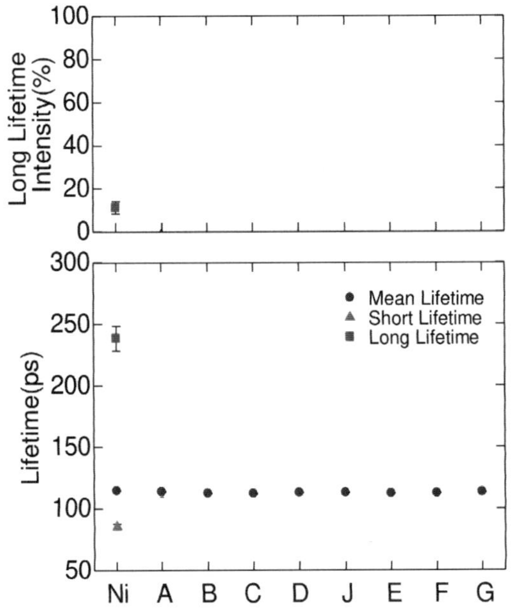

Figure 9. Positron annihilation lifetime of specimens irradiated with Ni ions at 573 K and 0.37 dpa.

Figures 8–10 show the positron annihilation lifetime of austenitic stainless steels and their model alloys irradiated with Ni ions at 573 K and 0.037, 0.37, and 6.8 dpa, respectively. Figures 11–12 show the Vickers hardness change of austenitic stainless steels and their model alloys. At 0.037 dpa, the mean lifetime increased after irradiation, but the lifetime spectra were not decomposed into two components. This is due to the formation of interstitial type dislocation loops or large stacking fault tetrahedra by the movement of vacancies. At 0.37 and 6.8 dpa, small vacancy clusters were detected only in pure Ni. In other alloys, the lifetime spectra were not decomposed into two components, and the large vacancy clusters could not be detected. These results are also identical to those obtained by neutron and electron irradiation experiments at 573 K [5,6]. Even at the high irradiation dose of 6.8 dpa, no voids were detected. For the formation of stable voids, an increase of the dislocation density by the growth of interstitial type dislocation loops is required (dislocation bias theory for void swelling [15]). The void swelling peak by ion irradiation is higher than that by neutron irradiation because of the high damage rate induced by ion irradiation [16]. In the present irradiation conditions, a temperature of 573 K may be too low for void swelling. Furthermore, after irradiation at 573 K, the hardness increased with the irradiation dose, but the change was small compared with that observed after irradiation at room temperature because of a lower density of the defect clusters. The low defect cluster density prevented the decomposition of the positron annihilation lifetime spectra of Ni-ion-irradiated austenitic stainless steels and their model alloys into two components.

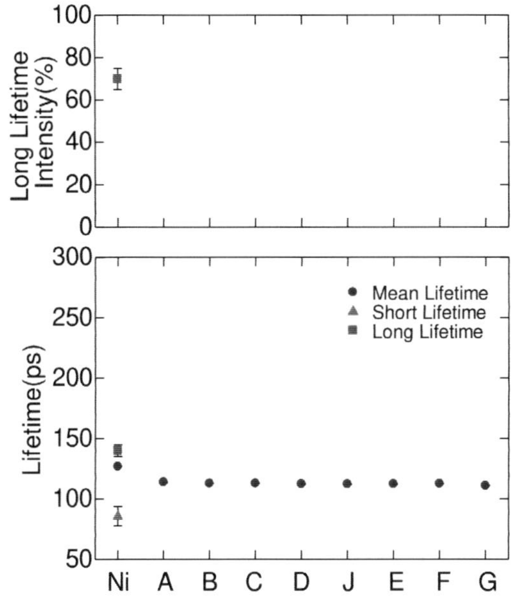

Figure 10. Positron annihilation lifetime of specimens irradiated with Ni ions at 573 K and 6.8 dpa.

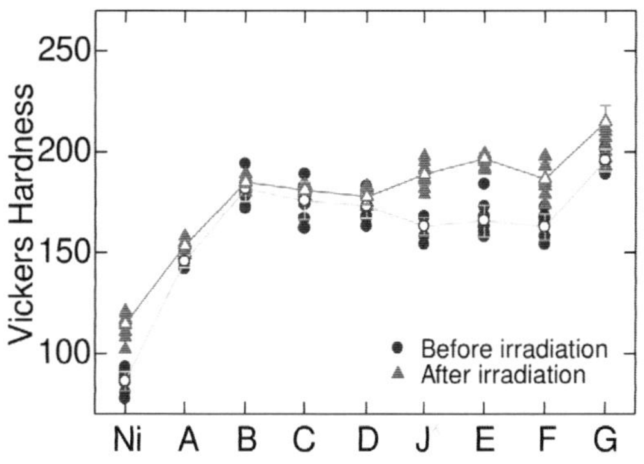

Figure 11. Vickers hardness of specimens irradiated with Ni ions at 573 K and 0.037 dpa.

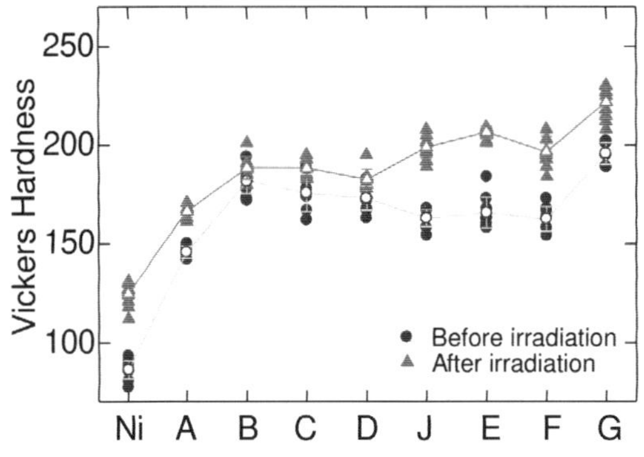

Figure 12. Vickers hardness of specimens irradiated with Ni ions at 573 K to 0.37 dpa.

4. Summary

Positron annihilation lifetime measurements and Vickers hardness tests of Ni-ion-irradiated stainless steels and their model alloys were performed. After irradiation at room temperature and at 573 K, the hardness increased with increasing the irradiation dose. However, the observed enhancement at 573 K was less intense. After irradiation at room temperature, a long lifetime of 150–180 ps was obtained, while after irradiation at 573 K the lifetime spectra were not decomposed into two components except for the Ni spectrum. These results are the same as those obtained by neutron and electron irradiation experiments [5,6].

References

[1] Katoh Y, Tanigawa H, Muroga T, Iwai T, and Kohyama A 1999 *J. Nucl. Mater.* **271&272** 115
[2] Cao XZ, Zhang P, Xu Q, Sato K, Tsuchida H, Cheng GD, Wu HB, Jiang XP, Yu RS, Wang BY and Wei L 2013 *J. Phys.* C **443** 012017
[3] Huang SS, Sato K, Xu Q, Yoshiie T, Tsuchida H, and Itoh A 2014 *J. Nucl. Mater.* **455** 122.
[4] Garner FA, Black CA and Edwards DJ 1997 *J. Nucl. Mater.* **245** 124
[5] Yoshiie T, Cao XZ, Sato K, Miyawaki K and Xu Q 2011 *J. Nucl. Mater.* **417** 968
[6] Yoshiie T, Cao XZ, Xu Q, Sato K and Troev TD 2009 *Phys. Status Sol.* C **6** 2333
[7] Kohno Y, Kohyama A and Kayano H 1992 *J. Nucl. Mater.* **191−194** 742
[8] Sawai T, Shiba K and Hishinuma A 1998 *J. Nucl. Mater.* **258−263** 1997
[9] Ziegler J F, Ziegler M D, and Biersack J P 2010 *Nucl. Instr. Met. Phys. Res.* B **268** 1818
[10] Olsen JV, Kirkegaard P, Pedersen NJ and Eldrup J 2007 *Phys. Stat. Sol.* C **4** 4004
[11] Ohkubo H, Tang Z, Nagai Y, Hasegawa M, Tawara T and Kiritani M 2003 *Mater. Sci, Eng.* A **350** 95
[12] Inamura M and Suzuki T 1990 *SEISAN-KENKYU/Report of the Institute of the Industrial Science, the University of Tokyo*, **42** 257 (Japanese)
[13] Kiritani M, Konno M, Yoshiie T and Kojima S 1987 *Mater. Sci. Forum* **15−18** 181
[14] Pokor C, Brechet Y, Dubuisson P, Massoud J P and Barbu A 2004 *J. Nucl. Mater.* **326** 19
[15] Bullough R, Eyre B L and Krishan K 1975 *Proc. Roy. Soc. London* A **346** 81
[16] Barton P J, Eyre B L and Stow D A 1977 *J. Nucl. Mater.* **67** 181

Electron irradiation-induced defects in Mo-diluted FeCrNi austenitic alloy during void swelling incubation

B Y Wang[1], E Y Lu[1], C X Zhang[1], Q Xu[2], S X Jin[1], P Zhang[1] and X Z Cao[1]

[1] Key Laboratory of Nuclear Radiation and Nuclear Energy Technology, Institute of High Energy Physics, CAS, Beijing, 100049, China

[2] Research Reactor Institute, Kyoto University, Kumatori-cho, Osaka 590-0494, Japan

E-mail: caoxzh@ihep.ac.cn

Abstract. The microstructural features and the effect of Mo addition in FeCrNi austenitic alloy during incubation period were investigated using positron annihilation technique and micro-Vickers Hardness. The electron irradiation, which could induce vacancy defects in material, was performed at room temperature up to the dose of 1.7×10^{-4} and 5×10^{-4} dpa, respectively. The defect concentration was estimated about 10^{-4}-10^{-7} though the standard trapping model. The added Mo atoms could trap vacancies to form Mo-vacancy complexes, which may restrain the migration and growth of vacancy defects during electron irradiation. In addition, the microstructural evolution during electron radiation resulted in hardening, while the added Mo might improve the hardening property of the alloy.

1. Introduction

Austenitic stainless steels are usually used as the structural material of nuclear reactors. Irradiation damage, such as void swelling and hardening, caused by prolonged energetic particle irradiation may affect the safety of nuclear reactors [1-2]. A transient stage or incubation period, which could determine the duration of the nuclear system, exists before steady void swelling occurs. Several theoretical and experimental analyses have been performed on the microstructural evolution during this period [3-6]. It is difficult to characterize the micro defects in this period by TEM for the reason that the defect size is usually lower than the resolution limit of TEM. Positron Annihilation Technique (PAT) is a suitable method to detect small-sized defects. It is well known that the defect structures of the incubation period relate with the irradiation temperature and dose. No microvoids would be found for the irradiation temperature above 473 K with the dose of 10^{-3} dpa in neutron irradiated Ti-modified 316 SS [7-8]. Meanwhile, the alloy elements may also affect the defect structures during the incubation period [1-2, 9-10]. As reported, undersize impurity atoms (P, Si) interacted with interstitial atoms, while oversized impurities (Ti, Nb) would interact with vacancies under irradiation in austenitic alloy [2]. These minor elements might induce the formation of stacking fault tetrahedral and precipitates, which could prevent the void growth [1, 9]. The added Mo could trap impurity atoms and reduce the number of dislocation loops under irradiation [10]. Additionally, Mo addition affect the formation of precipitates and crystal texture then enhance the mechanical properties and corrosion behaviour in different alloys [11-14]. As we known, the added Mo is the essential atom in FeCrNi austenitic stainless steel. This work was focus on the interaction of Mo addition with vacancy defects during incubation period.

In this study, FeCrNi and Mo-diluted model alloys were used to avoid the effect of other impurity elements. Electron irradiation was performed to generate vacancy defects. PATs, including Positron annihilation lifetime spectroscopy (PALS), Doppler broadening spectroscopy (DBS) and coincidence Doppler broadening (CDB), were used to investigate the microstructural evolution under electron irradiation and the effect of Mo addition on microstructures. Irradiation hardening was characterized by micro-Vickers Hardness.

2. Experimental details

FeCrNi alloy (17wt% Cr, 12wt% Ni, the balance Fe) and Mo-diluted alloy (17wt% Cr, 12wt% Ni, 2.2wt% Mo, the balance Fe) were made from a series of high purity metals (≥4N) by arc melting process at General Research Institute for Nonferrous Metals. Specimens with the size of 10×10×0.3 mm^3 were electron-chemical polished to have a mirror like surface, and then annealed at 1323 K for 2 h in vacuum (about 1×10^{-4} Pa).

Electron irradiation was performed by an electron linear accelerator of Research Reactor Institute, Kyoto University with an acceleration voltage of 8 MeV at less than 373 K. The damage rate was about 8×10^{-8} dpa/s and the irradiation dose was 1.7×10^{-4} dpa and 5×10^{-4} dpa, respectively.

PATs were performed to characterize the microstructural evolution of irradiated alloy at room temperature. PALS were carried out by a conventional fast-slow spectrometer with a time resolution of 197 ps (FWHM). A ^{22}Na positron source with the activity of ~10μCi was used. In order to reduce the statistical error, each spectrum accumulated about 2×10^6 coincidence events within 2 h. The spectra were decomposed into two components (τ_1 and τ_2) by LT9 after subtracting the source component and background. DBS were performed with a high purity germanium detector. The S parameter was defined as the ratio counts in the central energy region (511±0.76 keV) to the total counts of the spectrum (511±7.66 keV). CDB were performed to identify the chemical elements with two high purity germanium detectors placed 180° [15-16]. Each CDB spectrum was accumulated 1×10^7 counts to reduce the statistical error. Additionally, micro-Vickers Hardness was performed by a conventional apparatus with a load of 10 gF for 10 s in each measurement.

3. Results and discussion

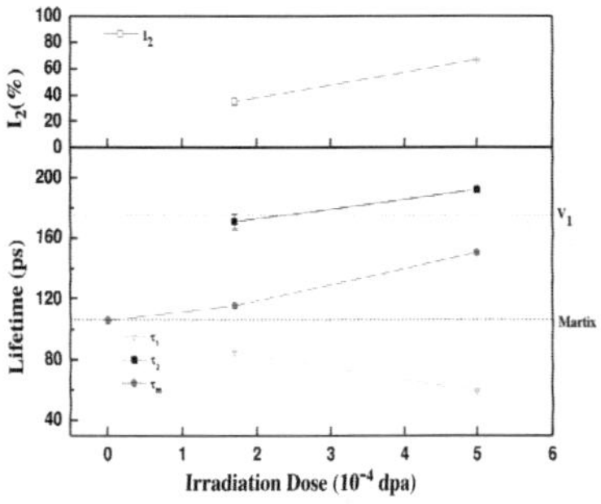

Figure 1. Dose dependence of the positron annihilation lifetime for electron irradiated FeCrNi model alloy.

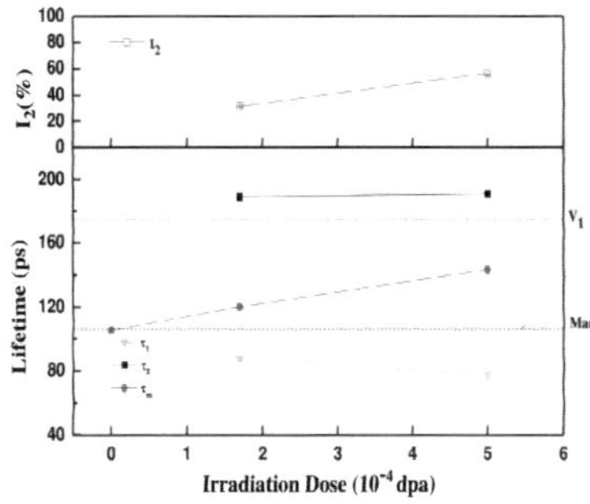

Figure 2. The positron annihilation lifetime for electron irradiated Mo-diluted alloy with different irradiation dose.

Figures 1 and 2 plot the dose dependence of the positron annihilation lifetime for electron irradiated FeCrNi and Mo-diluted alloy, respectively. τ_1, τ_2 and τ_m denote the short, long and mean positron lifetimes, respectively. Wang et al. had reported that the positron lifetime of well annealed Fe15Cr30Ni model alloy was 110±3 ps [17]. Meanwhile, the positron lifetime of matrix and mono-vacancy in AISI 316L was about 108 and175 ps [18], respectively. Dashed line of Matrix and V_1 denote the calculated positron lifetimes of the matrix and the mono-vacancy in pure Ni, which was similar with the experimental result in AISI 316L. The mean lifetimes for both unirradiated alloys are about 105 ps, as shown in Figures 1 and 2. In Figure 1, the long lifetime at low irradiation dose is 171 ± 5.1 ps, which is accord with the calculated mono-vacancy value (V_1). The result of higher irradiation dose is 189±5.1 ps, which is larger than V_1 but smaller than V_2. These results indicate that certain amount of small-size vacancy defects generated under electron irradiation. The increment of I_2 for both alloys indicates much more vacancy defects generated with elevated irradiation dose. However, I_2 in Mo-diluted alloy are slightly lower than that in FeCrNi alloy at the same irradiation condition. The added Mo might suppress the generation of vacancy defects during electron irradiation.

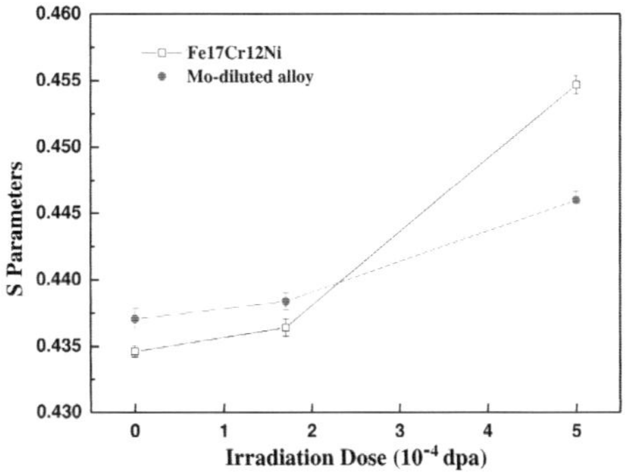

Figure 3. Dose dependence of the S parameters for FeCrNi and Mo-diluted alloys.

Figure 3. shows the S parameters of both alloys irradiated with different dose. The values of unirradiated alloys were attributed to the positron annihilated in matrix. The S parameters for both model alloys increased obviously with increasing irradiation dose, which means the increment of the defect concentration. According to the standard trapping model (STM) [19], the concentration of mono-vacancies can be estimated as:

$$C_V = \frac{\lambda_f(S-S_f)}{\mu_V(S_V-S)} \tag{1}$$

where λ_f is the annihilation rate in the defect free state; μ_V is the specific positron trapping rate into mono-vacancies. Because of the lattice structure and defects information is similar with pure Ni. Thus, the constant of Ni were chosen for our estimation (λ_f =9.1×10^9 s^{-1}; μ_V =2.2×10^{15} s^{-1} [20]); S_f and S_V are the S parameters characteristic of the positron annihilation from matrix and mono-vacancy-trapped state, respectively. S_f = 0.435 and 0.437 for FeCrNi and Mo-diluted alloy, respectively. As reported, S_V in pure Ni is equal to the saturated S value of cold rolled specimen (more than 10%)[21]. The saturated S value in plastic deformed Fe17Cr14.5Ni alloy has been identified about 0.455±0.001 at room temperature. Thus, S_V =0.455 was chosen in the calculation. The defect concentration was calculated about 10^{-7} for both alloys at lower irradiation dose, while 10^{-4} for FeCrNi alloy and 10^{-6} for Mo-diluted alloy at elevated irradiation dose.

Usually, point defects might be trapped by solute atoms because of the attribution of the solid solution effect [22]. Meanwhile, the oversized solute might reduce the migration of vacancies due to the high binding energy between the oversized atoms and vacancies [16]. In the present study, the added Mo, as oversized solute, existed in the alloy as solid solution state. The addition of Mo would trap vacancies and restrain the vacancy migration and growth. Therefore, swelling might be suppressed for the solid solution effect in Mo-diluted alloy.

CDB spectra could identify the chemical composition of the positron trapping site by measuring the momentum distribution of the electron annihilated with positron [15-16, 23-24]. As shown in Figure 4, the CDB ratio curves of irradiated specimens were normalized to the momentum distribution of FeCrNi alloy. One peak within the momentum region of $8\times10^{-3}m_0c$ <$|P_L|$<$18\times10^{-3}m_0c$ is observed in Figure 4 (b). The solute Mo atoms might trap vacancies to form Mo-vacancy complexes, which might change the chemical composition around vacancy sites compared to FeCrNi. Part of incident positrons, trapped by vacancies, annihilated with the characteristic core electron of Mo atoms. The effect of Mo-vacancy complexes induced the generation of the peak at high-momentum region.

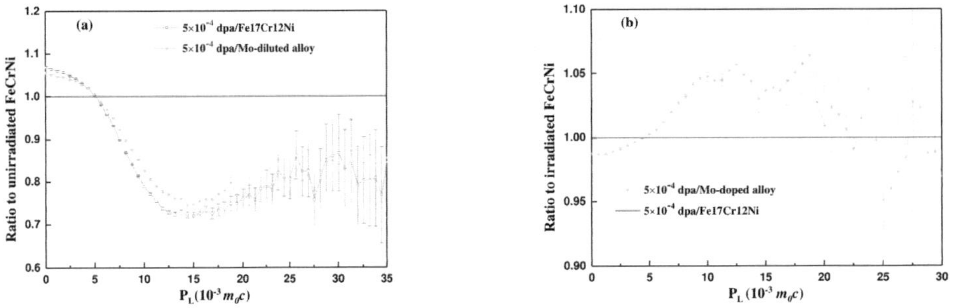

Figure 4. Ratio curves of CDB spectra in as-irradiated model alloy. (a) Normalized to the momentum distribution of as-annealed FeCrNi. (b) Comparison of two alloys at high irradiation dose. m_0 is the electron rest mass and c is the speed of light.

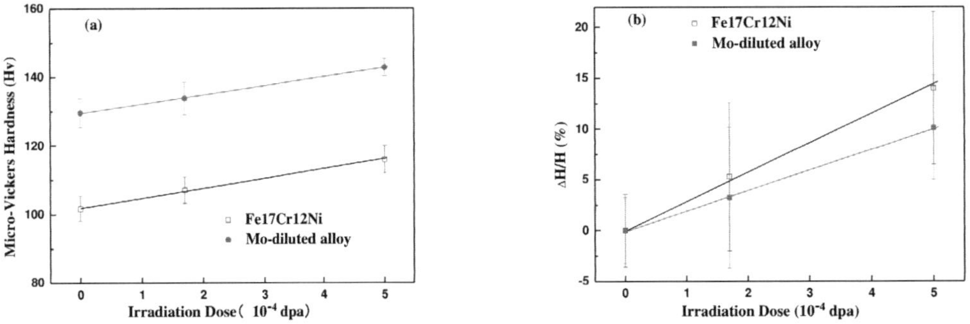

Figure 5. Dose dependence of micro-Vickers Hardness plot for different model alloys. (a) the value of micro-hardness, (b) ΔH/H parameters for both alloys with different radiation dose.

The hardness usually related with the microstructural features. The micro defects, such as dislocation loop, vacancy cluster or micro void and precipitates could induce hardening in materials [23, 25-26]. The micro-Vickers hardness value indicates that a significant difference exists between the two alloys, as shown in Figure 5. The increment of hardness for both alloys attribute to the vacancy defects induced by electron irradiation. $\Delta H/H=(H_{irra}-H_{unirra})/ H_{unirra}$, increase linearly for both alloys, as shown in Figure 5 (b). The slope of FeCrNi is larger than that of the Mo-diluted alloy. Irradiation hardening could be suppressed because of the Mo addition, which is consistent with the defect concentration. Additionally, the hardness value of Mo-diluted alloy was larger than that of the FeCrNi alloy at the same irradiation condition, as shown in Figure 5 (a). The solution strength induced by the added Mo atoms might be the main reason for this phenomenon [27-28].

4. Conclusion

PATs were performed to characterize the microstructural evolution in electron irradiated FeCrNi and Mo-diluted alloy as a function of dose dependence. The hardening process related with defect concentration was characterized by micro-Vickers Hardness. Certain amount of small-sized vacancies generated in the material after electron irradiation. The defect concentration was estimated about 10^{-4} for FeCrNi alloy and 10^{-6} for Mo-diluted alloy at elevated irradiation dose, while 10^{-7} for both alloys at lower dose. The diluted Mo could trap vacancy defects to form Mo-vacancy complexes and restrain the migration and growth of vacancy defects during electron irradiation. Hardening process was suppressed because of the Mo addition. Swelling would be suppressed in Mo-diluted FeCrNi austenitic alloy under irradiation.

Acknowledgements

This work is supported by the National Natural Science foundation of China 91226103 and 91026006.

References

[1] Yoshiie T, Sato K, Cao X, Xu Q, Horiki M and Troev T D 2012 *J. Nucl. Mater.* **429** 185
[2] Druzhkov A P, Perminov D A and Davletshin A E 2009 *J. Nucl. Mater.* **384** 56
[3] Garner F A 1984 *J. Nucl. Mater.* **122** 459
[4] Okita T, Kamada T and Sekimura N 2000 *J. Nucl. Mater.* **283** 220
[5] Okita T, Sato T, Sekimura N, Garner F A and Greenwood L R 2002 *J. Nucl. Mater.* **307** 322
[6] Brailsford A D and Bullough R 1972 *J. Nucl. Mater.* **44** 121
[7] Yoshiie T, Cao X Z, Xu Q, Sato K and Troev T D 2009 *Phys. Status Solidi* C **6** 2333
[8] Yoshiie T, Cao X Z, Sato K, Miyawaki K and Xu Q 2011 *J. Nucl. Mater.* **417** 968
[9] Horiki M, Yoshiie T, Huang S S, Sato K, Cao X Z, Xu Q and Troev T D 2013 *J. Nucl. Mater.* **442** S813
[10] Shigenaka N, Hashimoto T and Fuse M 1993 *J. Nucl. Mater.* **207** 46
[11] LaSalvia J C, Kim D K and Meyers M A 1996 *Mat. Sci. Eng.* A **206** 71
[12] Pardo A, Merino M C, Coy A E, Viejo F, Arrabal R and Matykina E 2008 *Corros Sci.* **50** 780
[13] Shi F, Qi Y and Liu C M 2011 *J. Mater. Sci. Technol.* **27** 1125
[14] Chun Y B, Hwang S K, Kim M H, Kwun S I and Chae S W 2001 *J. Nucl. Mater.* **295** 31
[15] Nagai Y, Takadate K, Tang Z, Ohkubo H, Sunaga H, Takizawa H and Hasegawa M 2003 *Phys. Rev. B.* **67** 224202
[16] AsokaKumar P, Alatalo M, Ghosh V J, Kruseman A C, Nielsen B and Lynn K G 1996 *Phys. Rev. Lett.* **77** 2097
[17] Wang T M, Wang B Y, Zhang S H and Doyama M 1992 *Phys. Stat. Solidi* A **134** 127
[18] Holzwarth U, Barbieri A, Hansen-Ilzhofer S, Schaaff P and Haaks M 2001 *Appl. Phys.* A **73** 467
[19] Druzhkov A P, Perminov D A and Arbuzov V L 2013 *J. Nucl. Mater.* **434** 198
[20] Druzhkov A P, Perminov D A and Pecherkina N L 2008 *Philos. Mag.* **88** 959
[21] Druzhkov A P, Arbuzov V L and Perminov D A 2012 *J. Nucl. Mater.* **421** 58
[22] Katoh Y and Kohyama A 1995 *Nucl. Instrum. Meth.* B **102** 12
[23] Nagai Y, Takadate K, Tang Z, Ohkubo H, Sunaga H, Takizawa H and Hasegawa M 2011 *Journal of Physics: Conference Series*. **265** 012007
[24] Hasegawa M, Nagai Y and Tang Z 2004 *32nd Course of the International School of Solid State Physics: Radiation Effects in Solids*. **July 19-28** 23
[25] Matsuoka H, Yamasaki T, Zheng Y J, Mitamura T, Terasawa M and Fukami T 2007 *Mat. Sci. Eng.* A **449** 790
[26] Liu X B, Wang R S, Jiang J, Wu Y C, Zhang C H, Ren A, Xu C L and Qian W J 2014 *J. Nucl. Mater.* **451** 249
[27] Tahir M I 1977 *Pak. J. Sci. Res.* **29** 39
[28] Franke O, Durst K and Goken M 2009 *J. Mater. Res.* **24** 1127

Journal of Physics: Conference Series **674** (2016) 012011 doi:10.1088/1742-6596/674/1/012011

Study of radiation damage in ODS steels by positron annihilation spectroscopy

I Bartošová[1,2], A Bouhaddane[1], M Dománková[3], V Slugeň[1], D Wall[4] and F A Selim[2]

[1] Institute of Nuclear and Physical Engineering, Slovak University of Technology in Bratislava, Ilkovičova 3, 812 19 Bratislava, Slovak Republic
[2] Department of Physics, Bowling Green State University, 104 Overman Hall, OH 43403, USA
[3] Department of Materials Science, Faculty of Materials Science and Technology in Trnava, Slovak University of Technology in Bratislava, Paulínska 16, 917 24 Trnava, Slovak Republic
[4] Nuclear Radiation Center, PO Box 641300, Washington State University, Pullman, WA 99164-1300, *USA*

E-mail: iveta.bartosova@stuba.sk

Abstract. Microstructure of various oxide-dispersion-strengthened (ODS) steels with 15% chromium content was studied in term of vacancy defects presence and their accumulation after defined irradiation treatment, respectively. Studied materials originated from Kyoto University and studied via IAEA collaborative project. Samples were characterized “as received” by positron annihilation lifetime spectroscopy and their microstructure was examined by transmission electron microscopy as well. Samples were afterwards irradiated in Washington State University Nuclear Radiation Center via a strong gamma source (6TBq). Damage induced by gamma irradiation was evaluated by positron lifetime measurements in emphasis on defect accumulation in the materials. We have demonstrated strong defect production induced by gamma irradiation which results from positron measurement data.

1. Introduction

Oxide dispersion strengthened steels (ODS) have been developed mainly for high temperature applications for advanced fusion reactor blanket components as well as the next generation fission reactor fuel cladding [1]. The microstructure of these steels is enhanced by dispersed particles of Y_2O_3 which influence dislocation movement and also serve as trapping sites for positrons due to high positron affinity of yttrium. Therefore investigation by means of positron annihilation lifetime spectroscopy (PALS) provides a wide range of information. Positron measurements are supported by transmission electron microscopy (TEM) which provides visual evaluation and overall parameters of the samples.

It is our understanding that the behaviour of ODS steels after gamma irradiation was not yet investigated. Traditionally gamma rays are utilized for food sterilization or irradiation of polymers and are not often utilized for material research of steels because only a really strong source and sufficient irradiation time can disrupt steel lattice in an extent that we can observe. This approach is therefore somewhat unexplored.

In the area of steel research gamma irradiation is mainly utilized to investigate its effect on the kinetics of carbon steels corrosion [2, 3] or for the production of corrosion related passive films on stainless steels, which present better behaviour and mechanical properties in comparison to un-irradiated samples [4] and provides a new insight into the role of competing oxide film kinetics [5]. There have been also some theoretical studies of gamma irradiation effect on the properties of ferritic reactor pressure vessel (RPV) steels. In [6] they employ MCBEND code to calculate gamma fluxes and energetic gamma ray displacement cross-sections calculated using either Baumann or Alexander methods. They also mention the mechanisms of gamma induced radiation damage of RPV steels. Also estimations of dpa cross sections have been done in Fe by several authors. In the work of Piñera et al. [7] several calculated estimates of dpa cross sections as a function of gamma ray energy of various research groups are compared. However vital data describing concrete gamma irradiation induced defect measurements are absent. We believe that gamma irradiation could be quite beneficial in comparison to other forms of irradiation. Our main concern with traditional neutron irradiation is the activity of the samples. Since we want to investigate the samples by PALS, the activated parasite cobalt peaks would interfere with the start signal of PALS equipment and therefore present false annihilation counts. Deconvolution of the spectra is possible, however quite high errors can be brought into the result. Also a 3-detector system can solve the problem with cobalt. On the other hand long measurement time and also the financial aspect seem as a draw-back. Also handling and waiting for the samples to cool down represents a significant problem. Therefore this novel approach could become an alternative to the traditional neutron irradiation.

Damage production by gamma rays has received less attention than damage by neutrons. Gamma rays create damage through creating fast electrons and it is by this interaction of electrons with atoms in the lattice that damage is produced. A ^{60}Co source emits two gamma rays of 1.17 and 1.33MeV. The main mechanism by which gamma rays of our range of energies induce damage is Compton scattering. Also photoelectric effect can contribute to damage formation. We calculated the maximum kinetic energy of the recoil atom, therefore achieving an elastic collision in a 180° angle, using equation 1 [8].

$$T_m = 2E(E + 2mc^2) * (Mc^2)^{-1} \quad (1)$$

E is the electron kinetic energy, mc^2 is the electron rest energy, M is the iron atom mass.

A displacement energy of 25-40eV is required for displacing an iron atom from the lattice. Literature states that electrons with energies above ≈0.4–0.7MeV (depending on crystallographic direction) will transfer sufficient energies to displace lattice atoms in iron [6]. For energies greater than ≈0.6 MeV, the maximum electron energy approaches the original gamma energy minus ≈0.25 MeV (in our case 0.92 and 1.08MeV). Solving equation 1 a maximum kinetic energy of about 68.5-87eV for the 1.17 and 1.33MeV energies was obtained; therefore applied gamma source is able to induce defects via production of fast electrons.

The aim of this work is to describe microstructural changes in materials lattice after gamma bombardment and determine the resistance of these ODS steels. We predict higher positron trapping due to higher concentration of defects caused by displacements in the lattice.

2. Experimental

2.1. TEM results

Three 15 wt%Cr ODS steels were delivered in form of cylinders. Heat treatment involved hot extrusion at 1150°C and subsequent air cooling. Cylinders were cut into smaller pieces using a band saw. The samples were then polished to high surface reflectance and ready for positron measurements.

For TEM further sample preparation was needed. Thin steel foil with thickness of 0.1mm was prepared by mechanical grinding. Afterwards targets with 3mm diameter were further thinned in Tenupol 5 device in a solution of: 300 ml HNO_3 + 700 ml CH_3OH, voltage 15V and electrolyte temperature 0°C. After cleansing in ethanol and drying, the sample was prepared for TEM. Investigations by TEM were carried out on JEOL 200CX electron microscope with accelerating voltage 200keV. Selected area electron diffraction (SAD) for identification of lattice parameters supplemented TEM analyses. Chemical composition of studied samples is summarized in table 1.

Figure 1 shows TEM observations of sample KOC5-3. The microstructure of all samples is very similar. The figure shows presence of polyhedric ferritic grains with relatively homogenic size. The mean size of ferritic grains for samples 4-3, 5-3 and 6-3 is d_{str} = 550±25 nm, d_{str} = 800±100 nm and d_{str}=650±40 nm, respectively. We also observed in figure 2 small globular particles, which show strong heterogeneity whether in size or distribution. Particle size varied from 2-300 nm however in all samples more than 60% of particles were 20-40 nm in diameter. SAD identified these particles to be Y_2O_3. Interestingly in sample 5-3 the Y_2O_3 phase was found in two modifications- bcc and also fcc. Also a presence of quasi globular particles was acknowledged in samples KOC4-3 and KOC5-3 which precipitate mainly on ferritic grain boundaries. Due to chemical composition and SAD results we believe the secondary phase is that of $M_{23}C_6$. Analyses also provided some evidence of precipitations interacting with Y_2O_3 particles at grain boundaries and dislocations as can be seen in figures 3 and 4.

Table 1. Chemical composition of studied steels KOC4-3, 5-3 and 6-3.

	KOC4-3	KOC5-3	KOC6-3
C	0.027	0.031	0.028
Si	0.03	0.03	0.03
Mn	0.02	0.02	0.01
P	0.005	0.005	0.005
S	0.002	0.002	0.002
Cr	15.33	15.15	15.41
W	1.9	1.9	1.8
Al	3.8	3.9	3.87
Ti	0.12	0.11	0.12
Y	0.26	0.26	0.26
O	0.14	0.14	0.14
N	0.009	0.005	0.007
Ar	0.006	0.005	0.005
Zr	0.32	0.58	-
Hf	-	-	0.59
Y_2O_3	0.33	0.33	0.33

From figure 4 it seems that small Y_2O_3 are localized in grain boundaries, which is appropriate since grain boundaries act as sinks. It is possible that they became localized during fabrication since hot extrusion at 1150°C and subsequent air cooling are part of the fabrication process. The area adjacent to the grain boundary seems relatively Y_2O_3 particle free. Y_2O_3 distort the surrounding crystal lattice which results in dislocation formation.

Positron annihilation lifetime spectroscopy was applied on as- received samples and after gamma irradiation. Positron lifetime measurements were carried out using ^{22}Na, with activity about 20μCi, deposited on a kapton foil. The source was sandwiched between two identical samples of steel. A minimum of $1.3x10^6$ counts was collected for each of the three runs conducted for individual samples. Positron lifetime data was evaluated using PATFIT-88 software. A detailed description of the

Figure 1. Bright field TEM image of sample KOC 5-3 in as- received state. Microstructure contains polyedric ferritic grains.

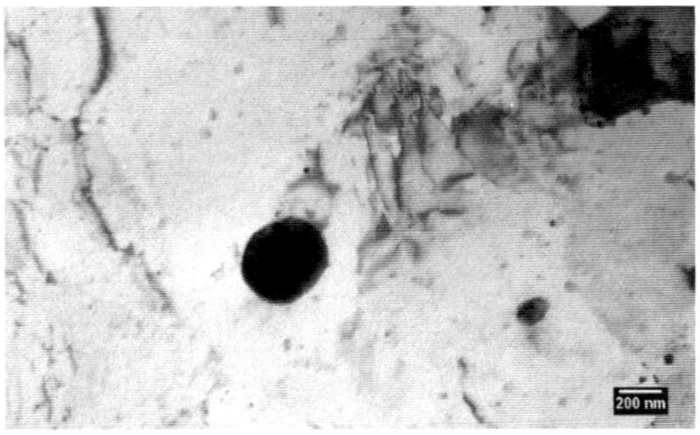

Figure 2. Detail on globular particle identified as Y_2O_3 in KOC4-3.

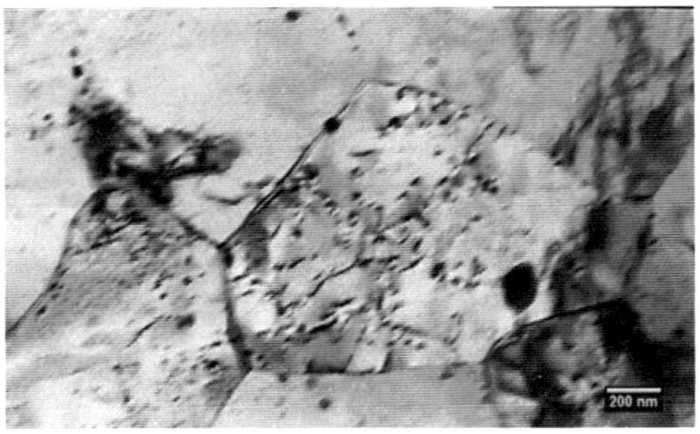

Figure 3. Interaction of dislocations with precipitates in KOC6-3.

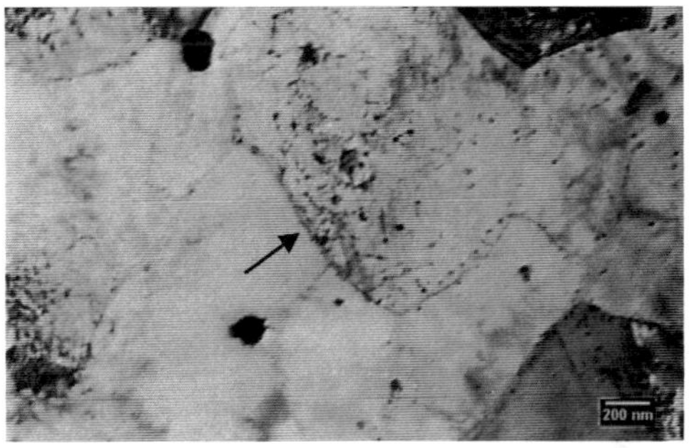

Figure 4. Area with low density of globular particles in ferritic matrix in KOC6-3. On the ferritic grain we can observe entrapment of dislocations around Y_2O_3 particles depicted by the arrow.

resolution function required for spectrum analysis used three Gaussian functions with intensities 80%, 10%, 10%, and appropriate relative shifts. The lifetime spectrum is analyzed as a sum of exponential decay components, convoluted with the Gaussians functions describing the spectrometer timing resolution using POSITRONFIT. Decay components due to annihilation in NaCl (≈430ps) and kapton foil (≈382ps) were subtracted in the procedure.

2.2. Irradiation details

Samples were irradiated at the Washington State University Nuclear Radiation Center using gamma rays generated by a ^{60}Co source. The ^{60}Co irradiator consists of a rod of radioactive Co metal which is stored in a pool of water at a depth of 7.6 meters. The ^{60}Co source material is adjacent to a 15 cm diameter aluminum tube which extends from above the surface of the pool down to the ^{60}Co source. The aluminum tube is sealed on the end which is at the bottom of the storage pool and is open to the atmosphere above the surface of the pool water; this allows samples to be placed in a basket to be lowered down into the aluminum tube and positioned adjacent to the ^{60}Co without having to immerse the samples in the storage pool water. The activity of the ^{60}Co source was 6.09 TBq and the irradiation was conducted continuously for 1742 hours. The distance from the ^{60}Co to the ODS samples was 9 cm. Sample temperatures were held at 30 ± 5° C during the irradiation.

3. Results

Figure 5 and 6 represent the results obtained from positron annihilation lifetime spectroscopy. The data was fitted by three exponential components where only the first two are interesting in characterization of studied materials. These lifetimes LT_1, LT_2 are shown in figure 5 and their relative intensities in figure 6. The average lifetime value was calculated using equation (2).

$$LT_{avg} = LT_1 I_1 + LT_2 I_2 \quad (2)$$

It is logical to assume that the shorter component LT_1 arises from annihilations of delocalized positrons. The longer component LT_2 represents contribution of the positrons trapped at one or more types of defects. The second component LT_2 for all the samples in as-received state is in the range 197-217 ps. This range corresponds to di-vacancies and suggests that positrons most likely get trapped and annihilate at defects of the same kind in all samples. The concentration of these defects in samples KOC4-3 and 5-3 is 55, 57% respectively; however in KOC6-3 the concentration is about 42% and I1 is much higher in comparison to materials KOC4-3, 5-3. Sample KOC6-3 contains much less defects in base state then the previous. By TEM technique we also asserted that small defects like dislocations are present in the samples.

It is generally believed that the dislocation line represents only a shallow positron trap. The positron diffuses until it finds a vacancy attached to the dislocation, where it afterwards annihilates. From TEM we saw that dislocations get trapped around Y_2O_3. Therefore we believe that the positron diffuses to a di-vacancy (from LT_2 value) attached to the dislocation which is surrounded by Y_2O_3 particles. The reduced positron lifetime LT_1 moves around 83-100ps and is in strong relation with LT_2 due to the fitting approach. The average lifetime in all samples is around 147±1ps before gamma irradiation.

After gamma irradiation we can observe major change in positron lifetime. Both lifetimes increased significantly. Lifetime LT_2 indicates formation of large clusters of vacancies in KOC4-3, 5-3 and vacancy clusters of about 4 vacancies in KOC6-3. The intensities also drastically shifted. After gamma irradiation I_2 decreased below 30% in all cases. Intensity I1 on the other hand increased above 70%. This indicates that gamma irradiation formed larger defects but with smaller intensity. Di-vacancies merged into larger defects forming vacancy clusters. Again, sample KOC6-3 seems to have fewer defects and is therefore more immune to gamma irradiation. The main difference in this steel is almost

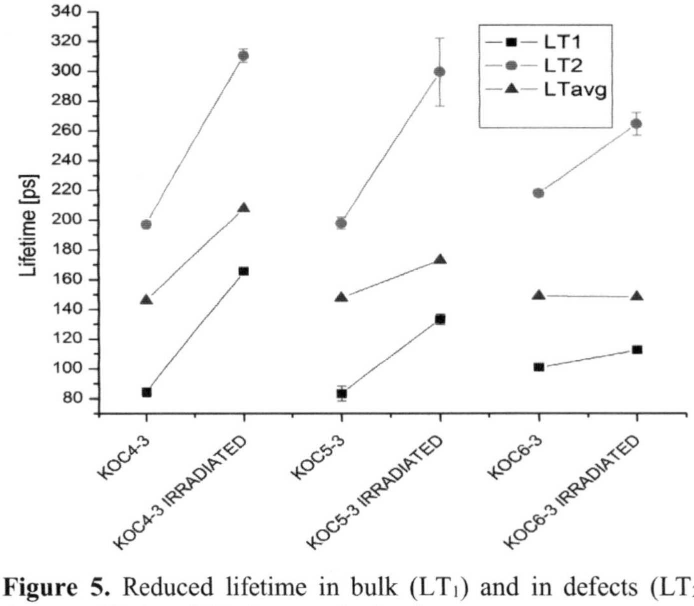

Figure 5. Reduced lifetime in bulk (LT_1) and in defects (LT_2). Average lifetime (LT_{avg}) was calculated.

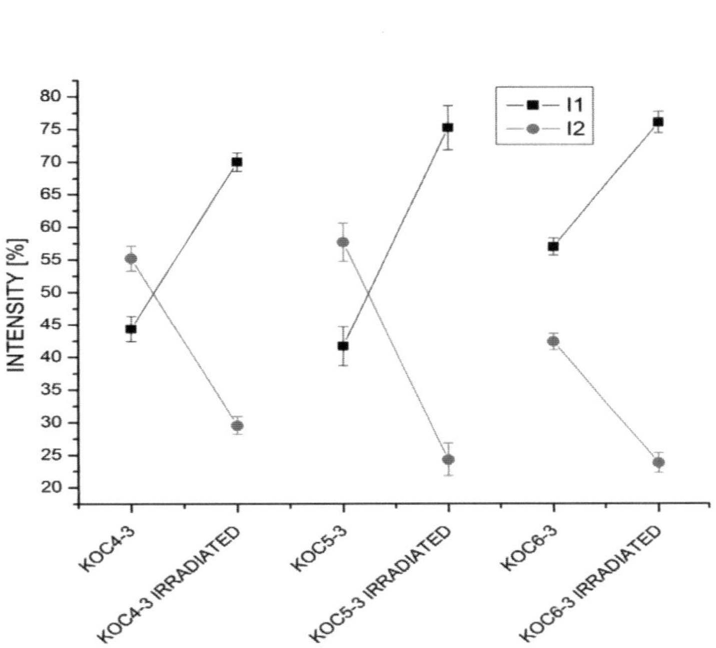

Figure 6. Intensities of particular lifetime components. Intensity I_1, I_2 corresponds to LT_1 and LT_2 respectively.

0.6 wt% content of hafnium and no zirconium additions. In [9] they implanted hafnium ions into foils of 9 wt%Cr ferritic steels to study the effect of hafnium on the grain boundary precipitation kinetics. It was found that the implantation of hafnium into the steel completely prohibited the formation of the common grain boundary $M_{23}C_6$ particles. In our case we can assume similar behaviour. Instead of $M_{23}C_6$ carbides hafnium forms MX type precipitates which increase the concentration of chromium in the matrix and is expected to improve the intergranular corrosion resistance of the material [10]. We do not state that all $M_{23}C_6$ formation was denied by hafnium, however we believe that hafnium plays a significant role in precipitation kinetics.

4. Conclusion

Three steels with different chemical content were investigated by transmission electron microscopy and positron annihilation spectroscopy. Samples designated KOC4-3 and KOC5-3 seem to have very similar properties as to defect type and concentration. Third sample KOC6-3 which is the only sample lacking zirconium and containing hafnium instead shows different behaviour as to initial microstructure and irradiation resistance after gamma irradiation. We attribute this improved resistance to hafnium which to some extent prohibits formation of $M_{23}C_6$ and forms hafnium carbides that have beneficial effect on the irradiation resistance.

We demonstrated a novel approach how to induce and investigate defects in steels. A high gamma irradiation source and appropriate irradiation time has proven to be sufficient by inducing major defects to the microstructure. Gamma irradiation does not interfere with positron measurements or TEM by undesirable side effects like in the case of neutron irradiation. In our future work we will also compare neutron irradiation and gamma irradiation on these samples. Nevertheless we believe that gamma irradiation of steels could be a helpful defect inducing method lacking the disadvantages of classical neutron irradiation.

References

[1] Kasada R, Lee S G, Isselin J, Lee J H, Omura T, Kimura A, Okuda T, Inoue M, Ukai S, Ohnuki S, Fujisawa T, Abe F 2011 *J.Nucl.Mat.* **417** 180

Journal of Physics: Conference Series **674** (2016) 012011 doi:10.1088/1742-6596/674/1/012011

[2] Fujita N, Matsuura CH, Kazuhiko S 2000 *Rad. Phys. and Chem.* **58** 139
[3] Daub K, Zhang X, Noel J J, Wren J C 2011 *Corrosion Science* **53** 11
[4] Capobianco G, Sandona G, Monetta T, Bellucci F 1993 *Corrosion Science* **35** 35
[5] Knapp Q W, Wren J C 2012 *Electrochimica Acta* **80** 90
[6] Knott J F, English C A, Weaver D R, Lidbury D P G 2005 *Inter. J. Press. Vess. and Pip.* **82** 929
[7] Piñera I, Cruz C M, Leyva A, Abreu Y, Cabal A E, Espen P V, Remortel N V 2014 *Nucl. Instr. and Meth. in Phys. Research* B **339** 1
[8] Piñera I, Cruz C M, Leyva A, Abreu Y 2007 *Nucleus*, **41** 39
[9] Yin Y F, Faulkner R G 2005 *Mat. Sc. and Tech.* **21** 1239
[10] Kim J K, Kim Y H, Lee J S, Kim K Y 2010 *Corr. Sc.* **52** 1847

Positron annihilation study of Fe-ion irradiated reactor pressure vessel model alloys

L Chen[1,4], Z C Li[2,5], H Schut[3] and N Sekimura[4]

[1]State Key Laboratory of New Ceramics and Fine Processing, School of Materials Science and Engineering, Tsinghua University, Beijing 100084, China
[2]Key Laboratory of Advanced Materials (MOE), School of Materials Science and Engineering, Tsinghua University, Beijing 100084, China
[3]Department of Radiation Science and Technology, Delft University of Technology, Mekelweg 15, 2629 JB Delft, the Netherlands
[4]Department of Nuclear Engineering and Management, The University of Tokyo, 7-3-1 Hongo Bunkyo-ku, Tokyo 113-8656, Japan

Email: zcli@tsinghua.edu.cn

Abstract. The degradation of reactor pressure vessel steels under irradiation, which results from the hardening and embrittlement caused by a high number density of nanometer scale damage, is of increasingly crucial concern for safe nuclear power plant operation and possible reactor lifetime prolongation. In this paper, the radiation damage in model alloys with increasing chemical complexity (Fe, Fe-Cu, Fe-Cu-Si, Fe-Cu-Ni and Fe-Cu-Ni-Mn) has been studied by Positron Annihilation Doppler Broadening spectroscopy after 1.5 MeV Fe-ion implantation at room temperature or high temperature (290 °C). It is found that the room temperature irradiation generally leads to the formation of vacancy-type defects in the Fe matrix. The high temperature irradiation exhibits an additional annealing effect for the radiation damage. Besides the Cu-rich clusters observed by the positron probe, the results show formation of vacancy-Mn complexes for implantation at low temperatures.

1. Introduction

The irradiation-induced embrittlement of nuclear reactor pressure vessel (RPV) steels is of great concern for the safety of the nuclear power plant (NPP) operation and possible lifetime prolongation since the first-generation reactors are approaching their initially designed operation lifetimes [1]. The basic task of nuclear reactor safety research is assessing the integrity of the RPV and its reliable lifetime prediction. Fundamental mechanisms associated with the embrittlement of RPV steels comprise the formation of solute clusters, matrix damage and grain boundary segregation, and have been studied in many comprehensive works [2-14]. However, there are still research issues that remain to be addressed, such as the mechanisms for the formation of vacancy-solute complexes.

The positron is a very suitable probe enabling sensitive detection of vacancy-containing damage in metals [15,16]. As anti-particle of the electron, the positron is trapped and annihilates at vacancy-type defect because at these sites the repelling Coulomb force due to the positively charged atomic nuclei is absent [17]. In addition, positrons annihilate sensitively at the solute clusters due to the difference in

[5] To whom any correspondence should be addressed.

positron affinity of the alloying atoms in comparison with the host Fe matrix [18,19]. The positron affinities found in the literature of the key alloying elements in this study are: -3.84 eV for Fe, -4.81 eV for Cu, -4.46 eV for Ni and -3.72 eV for Mn [18].

In the present paper, the Positron Annihilation Doppler Broadening (PADB) spectroscopy experiments were conducted to characterize the radiation damage in Fe-ion irradiated RPV model alloys, including Fe, Fe-Cu, Fe-Cu-Si, Fe-Cu-Ni and Fe-Cu-Ni-Mn. The results provide information on the formation of vacancy-type defects, Cu-rich clusters and vacancy-Mn complexes.

2. Materials and irradiation conditions

Five different series of RPV model alloys (in total eight types) were fabricated in the shape of a bar with a cross section of 10×10 mm^2 with growing chemical complexity, i.e. Fe (a purity of 99.95%), Fe-Cu, Fe-Cu-Si, Fe-Cu-Ni and Fe-Cu-Ni-Mn. The chemical composition of the alloys is provided in table 1. A final heat treatment at 900 °C for 30 minutes followed by water quenching was performed to release the stresses. From the bar specimens of size 10×10×2 mm^3 were prepared by wire cut electrical discharge machining. Finally the surfaces of all samples were polished to a mirror finish with metal polishing paper and diamond lapping.

Table 1. The chemical composition of the studied alloys.

Series	Concentration (wt.%)			
	Cu	Ni	Mn	Si
Fe	≤ 0.0001	≤ 0.0012	≤ 0.00025	≤ 0.0004
Fe-Cu	0.5	-	-	-
Fe-Cu-Si	0.5	-	-	0.1
	0.5	-	-	0.2
Fe-Cu-Ni	0.5	0.6	-	-
	0.5	0.8	-	-
Fe-Cu-Ni-Mn	0.5	0.6	1.4	-
	0.5	0.8	1.4	-

The eight types of specimens have been irradiated at room temperature (RT) with 1.5 MeV Fe ions at the 320 kV implantation platform at the Institute of Modern Physics, Chinese Academy of Sciences, to doses of 0.03 dpa, 0.06 dpa and 0.14 dpa (0.1 dpa corresponding to about 40 years of radiation and the theoretical end-of-life of the NPPs). In addition, five types of specimens, Fe, Fe-0.5%Cu, Fe-0.5%Cu-0.2%Si, Fe-0.5%Cu-0.8%Ni and Fe-0.5%Cu-0.8%Ni-1.4%Mn were selected to be irradiated at 290±10 °C (high temperature, denoted as "HT") to a dose of 0.14 dpa, as RPVs in service operate at about 290±30 °C. In the present paper, Fe^{10+} was used for the implantation (i.e. ion irradiation) and the implantation flux is 1.2×10^{11} cm^{-2}s^{-1}. The damage level, dpa associated with the ion irradiation was determined by SRIM. The quick Kinchin and Pease option with a displacement energy of 40 eV and a lattice binding energy of 0 was run, and then the number of displacements was calculated according to the damage energy to be most consistent with the internationally-recognized standard NRT dpa, as recommended by Stoller R E et al. [20]. Then the dose rate D in dpa/s can be calculated as follows:

$$D = \frac{F N^{id}}{\rho} \quad (1)$$

where F is the ion flux, N^{id} the number of displacements per ion and per depth and ρ the target's atom density of 8.48×10^{22} atoms/cm^3. This leads to a dose rate of 1.5×10^{-4} dpa/s (peak value) for the Fe-ion implantation in this present paper.

3. Measurement and analysis

The PADB measurements were carried out at the Delft ^{22}Na-source based Variable Energy Positron beam facility (VEP). This PADB spectroscopy provides not only information about open volume defects but also about the chemical environment of the positron annihilation site by measuring the momentum distribution of the electron-positron annihilated pair.

The momentum of the electron-positron pair prior to annihilation produces a Doppler shift $\Delta E = \pm p_L c/2$ on the energy of the two annihilation photons [21,22]. Here p_L is the longitudinal component of the momentum along the direction of the gamma ray emission and c is the speed of light. The overall effect is observed as a Doppler broadening the 511 keV photo-peak in the annihilation gamma spectrum.

SRIM calculations show that the damage caused by the implantation of the 1.5 MeV Fe ions is below 800 nm with a maximum at a depth of about 450 nm. For the PADB experiments, the maximum positron implantation energy is 25 keV which corresponds to a positron mean implantation depth of about 900 nm, thus covering the sub-surface damage region.

The PADB spectra were currently measured with one gamma detector in order to get an overview of the irradiation damage along the entire Fe-ion implantation depth. Two Doppler parameters, *S* and *W* were extracted from each spectrum. The *S* parameter is the ratio of low-momentum region in the PADB spectrum to the total region, and the *W* parameter is the corresponding fraction of high-momentum annihilations. Besides showing *S* (*W*) data versus positron implantation depth (*d*), it is convenient to represent the data in *S*-*W* plots with energy (or mean implantation depth) as running parameter).

4. Results and discussion

As previously mentioned, in the present systems, the positrons are sensitively trapped at vacancy-type sites (i.e. vacancy-type defects and vacancy-solute complexes) and in nano-sized solute cluster with high (absolute value) positron affinity.

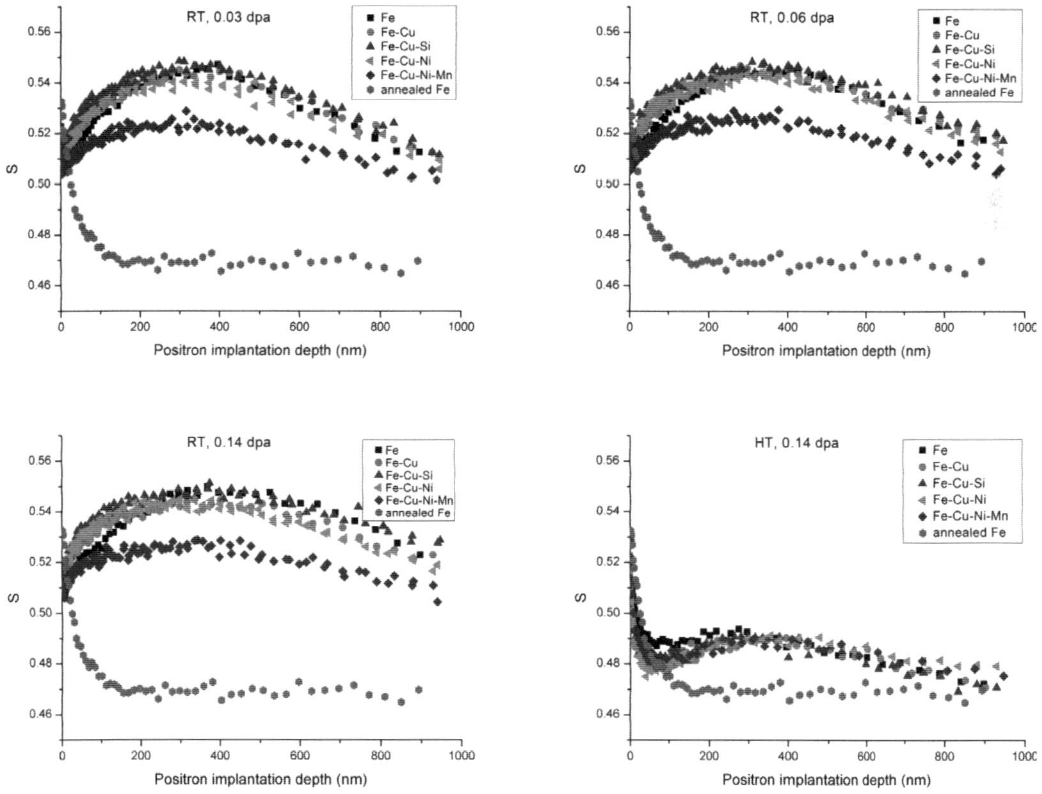

Figure 1. The *S* parameter as a function of positron mean implantation depth of all samples for the different implantation doses (0.03 dpa, 0.06 dpa and 0.14 dpa) at RT and for 0.14 dpa at HT.

Journal of Physics: Conference Series **674** (2016) 012012 doi:10.1088/1742-6596/674/1/012012

Figure 1 shows the *S* parameter as a function of positron mean implantation depth of all samples for the different implantation doses (0.03 dpa, 0.06 dpa and 0.14 dpa) at RT and for 0.14 dpa at HT. The S curves approaching an *S* value of 0.47 are obtained for an annealed Fe sample (an extra non-irradiated Fe which is annealed after surface treatment). For all samples the behavior of the *W* parameter is roughly the inverse (high *S* - low *W*) and is therefore not shown. The data of the RT implantations show virtually no dose dependence indicating positron saturation trapping. Furthermore it can be seen that for alloys containing additional Mn the *S* parameter in the implantation range is significantly lower. For the 0.14 dpa implantation at 290 °C the *S* parameter in the implantation region has reduced to a value of about 0.49 for all samples.

Figure 2 shows the *S-W* plots obtained for RT irradiated Fe, HT irradiated Fe and non-irradiated Fe as well as the annealed Fe. The arrows indicate the direction of increasing positron implantation energy or implantation depth. Starting from the surface, for the RT data the *S*,*W* points move into a direction of a cluster point that is also present in the non-irradiated Fe. Halfway, the points move into the direction of a line that connects the characteristic (*S*,*W*) cluster point for Fe with the (*S*,*W*) anchor point with highest *S* value. Having reached this line the data follows the line up to the highest *S* value where afterwards they follow the same line but now in the direction of the (*S*,*W*) cluster point for Fe. As the Fe used in this work does not contain solute elements (i.e. the content of solute elements is low enough to be neglected here), this round-trip curvature following the specific *S*,*W* anchor points hints at the formation of vacancy-type defect with a depth dependent defect concentration.

A closer inspection of the *S*,*W* data obtained at the non-irradiated Fe shows an anchor point in the near surface region (< 300 nm). This is somewhat unexpected, because in non-irradiated Fe the data should lie a straight line, as is found for the annealed Fe. The possible cause for this anchor point (which is below the defect - bulk line) is that the mechanical surface polishing has introduced near surface defects likely consisting of dislocation loops and vacancies. The figure shows that the anchor point for the HT irradiated Fe is close to the one just described for the non-irradiated Fe. The lower *S* value supposedly is caused by the additional annealing effect of the HT irradiation which effectively enables the recombination of vacancies with self-interstitial atoms [23].

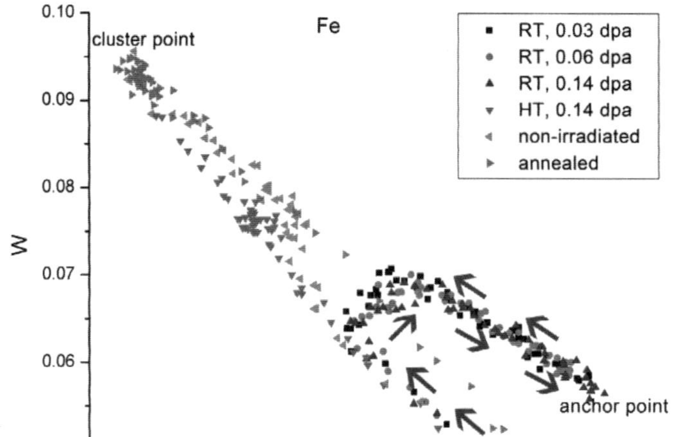

Figure 2. Results of PADB data analysis for Fe under irradiation conditions with variations in temperature and in dose as well as non-irradiated Fe and annealed Fe.

Figure 3 shows the *S-W* plots for the Fe-Cu alloy containing 0.5% of Cu. We can see that Fe-ion irradiation again leads to the round-trip curvature in the *S-W* plot for all RT irradiated Fe-0.5%Cu alloys. It is reported that the *W* for elemental Cu is larger than that for elemental Fe [22] while the *S* parameter is only slightly lower. The effect of Cu is accordingly seen in the *S*-*W* plots in which the (*S*,*W*) anchor points belonging to the implantation region all are shifted upwards (increased W value). Also the HT data shows the round-trip behavior, but it is now located at low *S*, high *W* corner of the figure.

Compared to the RT data, the HT irradiated shows a significant lower peak value of *S* and its value is comparable to the one obtained for the HT irradiated Fe. The *W* parameter however shows a relatively larger increase. This observation can be interpreted as follows. It is well known that the Cu diffusivity depends on the vacancy concentration [12]. The annealing effect of HT irradiation in the end reduces the final vacancy concentration, but during the implantation process the available vacancies may assist the Cu diffusion leading to high number density of Cu clusters. The first effect makes the data shift into the left top corner of the graph while the latter further increases *W* only.

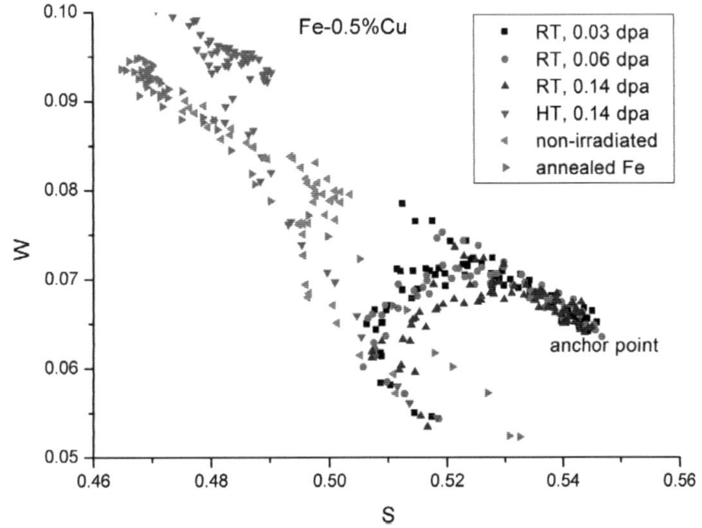

Figure 3. Results of PADB data analysis for irradiated and non-irradiated Fe-0.5%Cu alloys as well as annealed Fe.

The (*S*,*W*) points with maximum *S* for all RT irradiated alloys are presented in figure 4. These points are located at the end of the line which connects the (*S*,*W*) cluster point for Fe and (*S*,*W*) anchor point for the implantation defects in the *S*-*W* plots. It is clear that under Fe-ion irradiation, the Mn-containing samples (Fe-Cu-Ni-Mn) give lower values of *S* compared with the other samples. This is even the case when comparing the 0.14 dpa irradiated Fe-Cu-Ni-Mn with the 0.03 dpa irradiated Mn-free alloys. As mentioned before, the positron affinity of Mn is lower than that of matrix Fe. Therefore, a possible explanation is that in RT irradiated Fe-Cu-Ni-Mn alloys the vacancy-Mn complexes form efficiently, thus decreasing the content of vacancies and/or vacancy clusters. This is then unfavorable for the formation of Cu-rich clusters due to lower Cu diffusivity. In addition, the low positron affinity of Mn impedes positron trapping at the vacancy-Mn complexes. All of this leads to the decrease of *S* for the Mn-containing model alloys. This explanation is in agreement with the possible mechanism suggested for the effect of Mn on vacancy clusters and Cu-rich clusters: additional Mn reduces the overall vacancy diffusion, vacancy clustering, and the corresponding radiation enhanced diffusion of Cu necessary for the formation of Cu-rich clusters due to a strong interaction between Mn and individual point defects [24].

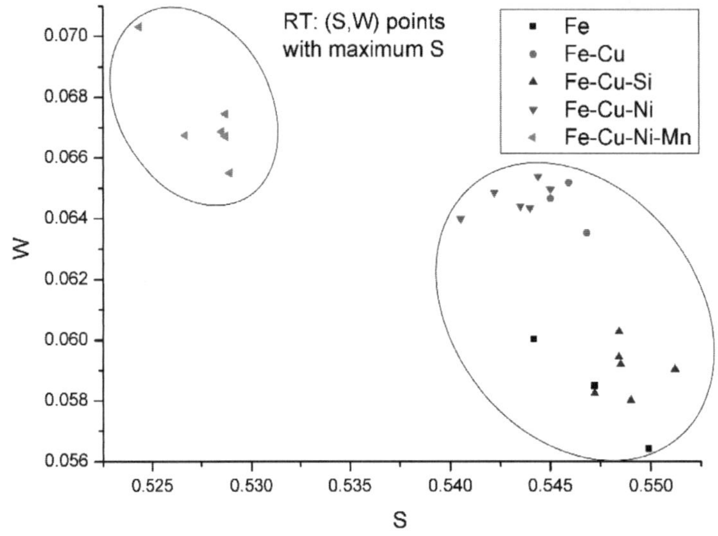

Figure 4. The (S,W) points with maximum S for all RT irradiated samples to doses of 0.03 dpa, 0.06 dpa and 0.14 dpa.

Next, the effect of Mn under HT irradiation was analyzed. Figure 5 shows the S-W plots for HT irradiated samples. It is observed that under HT irradiation, compared to the Mn-free alloys, the Mn-containing alloy has a lower W at maximum S for the round-trip curvature in the S-W plot. As reported, the W for elemental Mn is the lowest among those for Fe, Cu, Ni and Mn [22]. Then the low W observed in Fe-0.5%Cu-0.8%Ni-1.4%Mn alloy can be explained by the enrichment of Mn in Cu-rich clusters. In HT irradiated Fe-0.5%Cu-0.8%Ni-1.4%Mn alloys, there is no observation of the evidence for the formation of vacancy-Mn complexes, in terms of the lowering of the S value for the round-trip curvature part in the S-W plot. Therefore, it is reasonable to conclude that low temperature favors the formation of vacancy-Mn complexes during irradiation. Furthermore, it has been indicated that Mn-rich clusters are promoted at lower irradiation temperatures [25]. The consistency of these two microstructures with respect to the irradiation temperature supports the reported mechanism that vacancy-Mn complexes may be precursors to the formation of Mn-rich clusters [22].

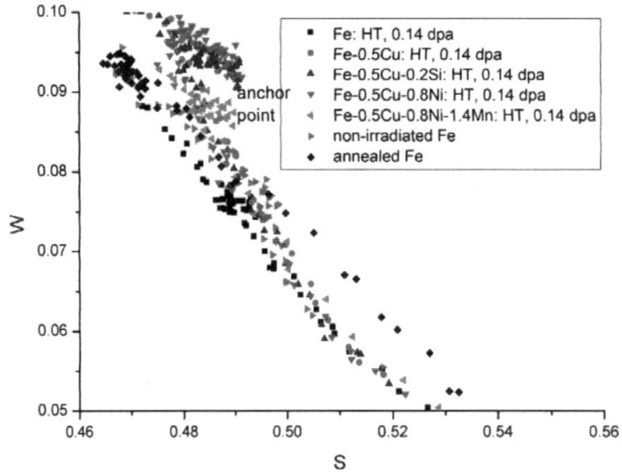

Figure 5. Results of PADB data analysis for HT irradiated Fe and Fe- alloys as well as non-irradiated Fe and annealed Fe.

5. Conclusion
The results of the PADB experiments conducted in this study are used to characterize the positron-sensitive nano features produced in the RPV model alloys (Fe, Fe-Cu, Fe-Cu-Si, Fe-Cu-Ni and Fe-Cu-Ni-Mn) subjected to 1.5 MeV Fe-ion irradiation. It is observed that vacancy-type defects are formed in RT and HT irradiated Fe and that the higher temperature leads to an additional annealing effect. The PADB measurements also show the formation of Cu-rich clusters in ion irradiated Fe-Cu alloys and HT irradiated Fe-Cu-Si, Fe-Cu-Ni and Fe-Cu-Ni-Mn alloys. The behaviors of Mn under RT irradiation and under HT irradiation are found to be different. It is shown that in Fe-Cu-Ni-Mn alloys, vacancy-Mn complexes are formed under RT irradiation, but not formed under HT irradiation. This indicates that during irradiation, low temperature favors the formation of vacancy-Mn complexes.

Acknowledgments
This work was supported by the National Science and Technology Major Project (under Grant 2011ZX06004-002), and the National Natural Science Foundation of China (under Grant 61176003).

References
[1] Odette G R and Nanstad R K 2009 *JOM* **61** 17-23
[2] Lambrecht M and Almazouzi A 2011 *J. Phys.: Conf. Ser.* **265** 012009
[3] Styman P D, Hyde J M, Wilford K, Morley A and Smith G D W 2012 *Prog. Nucl. Energ.* **57** 86-92
[4] Toyama T, Nagai Y, Tang Z, Hasegawa M, Almazouzi A, van Walle E and Gerard R 2007 *Acta Mater.* **55** 6852-60
[5] Meslin E, Radiguet B and Loyer-Prost M 2013 *Acta Mater.* **61** 6246-54
[6] Li Z C, Abe H and Sekimura N 2007 *J. Nucl. Mater.* **362** 87-92
[7] Hernández-Mayoral M and Gómez-Briceño D 2010 *J. Nucl. Mater.* **399** 146-53
[8] Slugen V, Kögel G, Sperr P and Triftshäuser W 2002 *J. Nucl. Mater.* **302** 89-95
[9] Soisson F, Barbu A and Martin G 1996 *Acta Mater.* **44** 3789-800
[10] Stoller R E 2000 *Nucl. Eng. Des.* **195** 129-36
[11] Li Z C and Chen L 2014 *Acta Metall. Sin.* **50** 1285-93
[12] Christien F and Barbu A 2004 *J. Nucl. Mater.* **324** 90-6
[13] Schmauder S and Binkele P 2002 *Comp. Mater. Sci.* **24** 42-53
[14] Zhang Q, Li Z C, Lin C, Liu B X and Ma E 2000 *J. Appl. Phys.* **87** 4147-52
[15] Nagai Y, Takadate K, Tang Z, Ohkubo H, Sunaga H, Takizawa H and Hasegawa M 2003 *Phys. Rev.* B **67** 224202
[16] Hu Z, Li Z C, Zhou Z, Shi C Q, Schut H and Pappas K 2014 *J. Phys.: Conf. Ser.* **505** 012014
[17] Eldrup M and Singh B N 1997 *J. Nucl. Mater.* **251** 132-8
[18] Puska M J and Nieminen R M 1994 *Rev. Mod. Phys.* **66** 841-97
[19] Brauer G, Puska M J, Sob M and Korhonen T 1995 *Nucl. Eng. Des.* **158** 149-56
[20] Stoller R E, Toloczko M B, Was G S, Certain A G, Dwaraknath S and Garner F A 2013 *Nucl. Instrum. Methods Phys. Res. Sect.* B **310** 75-80
[21] Lambrecht M and Almazouzi A 2009 *J. Nucl. Mater.* **385** 334-8
[22] Glade S C, Wirth B D, Odette G R and Asoka-Kumar P 2006 *J. Nucl. Mater.* **351** 197-208
[23] Liu J, Yao H J, Sun Y M, Duan J L, Hou M D, Mo D, Wang Z G, Jin Y F, Abe H, Li Z C and Sekimura N 2006 *Nucl. Instrum. Methods Phys. Res. Sect.* B **245** 126-9
[24] Glade S C, Wirth B D, Odette G R, Asoka-Kumar P, Sterne P A and Howell R H 2005 *Philos. Mag.* **85** 629-39
[25] Lucas G E 2010 *J. Nucl. Mater.* **407** 59-69

Positron annihilation in transparent ceramics

P Husband[1], I Bartošová[2], V Slugeň[2], F A Selim[1,3,*]

[1]Department of Physics and Astronomy, Bowling Green State University, Bowling Green, OH 43403, USA
[2]Institute of Nuclear and Physical Engineering, Slovak University of Technology in Bratislava, Ilkovičova 3, 812 19 Bratislava, Slovak Republic
[3]Center for Photochemical Sciences, Bowling Green State University, Bowling Green, OH 43403, USA

E-mail: faselim@bgsu.edu

Abstract: Transparent ceramics are emerging as excellent candidates for many photonic applications including laser, scintillation and illumination. However achieving perfect transparency is essential in these applications and requires high technology processing and complete understanding for the ceramic microstructure and its effect on the optical properties. Positron annihilation spectroscopy (PAS) is the perfect tool to study porosity and defects. It has been applied to investigate many ceramic structures; and transparent ceramics field may be greatly advanced by applying PAS. In this work positron lifetime (PLT) measurements were carried out in parallel with optical studies on yttrium aluminum garnet transparent ceramics in order to gain an understanding for their structure at the atomic level and its effect on the transparency and light scattering. The study confirmed that PAS can provide useful information on their microstructure and guide the technology of manufacturing and advancing transparent ceramics.

1. Introduction

Transparent ceramics are optical materials formed by consolidation of nanocrystalline precursors into a fully dense solid [1]. They feature good mechanical strength, high melting points and good transmission in the IR spectrum. They have some advantages over single crystals as they can be molded and shaped as needed for any particular application. Moreover, manufacturing transparent ceramics does not require the same amount of time or complex facilities that single crystals require [2]. Cubic crystal structure is the best choice for transparent ceramics to minimize light scattering and lanthanide aluminum garnets-in particular- were shown to be promising transparent ceramics for laser and scintillation applications [1].
These potential applications require perfect transparency and minimum light scattering. However the presence of grains and grain boundaries tend to increase light scattering. Therefore it is critical to study the microstructure of transparent ceramics and reveal their grain sizes and distributions. The focus of this work is to study the transparency and microstructure of undoped $Y_3Al_5O_{12}$, yttrium aluminum garnet (YAG) and Ce doped YAG transparent ceramics. This will be done by combining positron lifetime (PLT) spectroscopy with optical transmission measurements. Positron annihilation spectroscopy is a powerful tool for investigating defects and open volumes in solids and polymers. It has been successfully applied to

study the microstructure of a number of ceramics [3-13], as large fraction of positrons often annihilate at the grain boundaries providing information about their size and character.

The basics of PLT spectroscopy of defects and microstructures can be explained as follows [14, 15]. When positron enters a medium, it thermalizes in about 10 ps, then diffuses until it annihilates with an electron emitting two 511 keV photons. In homogeneous defect-free matter, all positrons annihilate in the bulk with characteristic bulk lifetime τ_B. In the presence of defects or open volumes, the positron may become trapped in open-volume defects (monovacancies, larger vacancy clusters, grain boundaries or open volumes). This is because the positron is repelled by the positively charged ion cores. Hence, defects with missing atoms or a reduced density of ions provide an attractive potential for positrons. Positron trapping at defects lead to longer lifetime. The positron lifetime is the difference between time of creation of the positron (usually marked using a time-coincident gamma-ray in case of radioactive source or using the beginning of a pulse in case of pulsed positron beam) and the time of annihilation, marked by detection of one of the 511 keV annihilation radiations. This gives the lifetime of a single positron, and one typically accumulates a spectrum of lifetimes over a period of hours. The annihilation rate λ, reciprocal to the mean positron lifetime τ, is given by [15]:

$$\lambda = \frac{1}{\tau} = \pi\, r_o\, c^2 \int \left|\psi(r)\right|^2 n(r)\, \gamma\, dr \qquad (1)$$

in which r_o is the classical electron radius, c is the speed of light, r is the position vector, $|\Psi(r)|^2$ is the positron probability density, n(r) is the electron density and γ is an enhancement factor representing an increase in the local electron density due to the attractive coulomb interaction between the positron and nearby electrons. Since the positron lifetime is a function of the electron density at the annihilation site, lifetime measurements can discriminate among the different lattice locations at which the positron annihilates, and consequently give information about defects and open volumes. In a defect-free material, positrons form Bloch states and annihilate as free particles with a well-defined "bulk" annihilation rate. This leads to a single exponential decay [15] in the lifetime spectrum with a characteristic "bulk" lifetime. Positrons annihilating in different defects will have different characteristic lifetimes τ_i. A general lifetime spectrum N(t) will be a sum of components corresponding to each annihilation site:

$$N(t) = \sum_{i=1}^{k+1} \frac{I_i}{\tau_i} \exp\left(\frac{-t}{\tau_i}\right) \qquad (2)$$

in which k+1 is the number of time components in the spectrum relating to k defects and for positron annihilation in the bulk; τ_i and I_i are the lifetime and intensity of the ith component in the spectrum. The spectrum is fitted with Eq. (2) after convoluting the sum with the time-resolution function of the spectrometer. The fitted lifetimes give information about defect/open volume sizes and characteristics and the intensities determine their concentrations [14,15].

In porous materials with large open volumes, the positron may first form a positronium with an electron, then annihilate based on the relative spin orientation between the electron and positron. The positron lifetimes in vacuum for para-positronium (singlet state) and ortho-positronium (triplet state) are 125 ps and 142 ns respectively. In matter, the pick-off annihilation process [15] takes place leading to few nanoseconds lifetimes.

2. Experimental details

Transparent ceramics of undoped YAG and Ce doped YAG samples were obtained from Baikowski/ Konoshima Inc. They are round samples of 1 cm diameter and 1 mm thickness with very good polished surfaces. Figure 1 shows an image for the undoped and Ce doped YAG samples with their names written underneath to demonstrate their good transparency.

Figure 1. A photo of undoped YAG and Ce doped YAG transparent ceramics samples

Positron lifetime spectra were measured using a conventional fast-fast time coincidence spectrometer with two BaF_2 detectors in a collinear geometry and with time resolution of ~200 ps [16]. ^{22}Na with about 20μCi activity was used as a positron source; it was made by depositing NaCl on a 0.8 microns kapton foil. The source was sandwiched between two identical samples. Na-22 emits a photon with the energy of 1.274 MeV in coincidence with each positron. This photon can be used to provide a start signal for PLT measurements. One of the two 511 keV annihilation photons is used to provide the stop signal. Each PLT spectrum contained at least $1.3x10^6$ counts and multiple runs were made for better statistics. The spectra were analyzed using PATFIT'88 software [16]. A detailed description of the resolution function required for spectrum analysis in PATFIT'88 used three Gaussian functions. The lifetime spectrum was then analyzed as a sum of exponential decay components with the Gaussians functions describing the spectrometer timing resolution. Decay components due to annihilation in NaCl (≈430ps) and kapton foil (≈382ps) were subtracted from each spectrum. These values have been previously determined for the setup using annealed defect free aluminum single crystals.

Optical transmission measurements were conducted from 190-1100 nm using a Perkin Elmer UV-VIS-NIR Spectrometer [17]. For comparison, measurements were also made on undoped and Ce doped YAG single crystals of the same size and thickness.

3. Results and discussion

All PLT spectra were well fitted to two lifetime components τ_1 and τ_2 and their intensities I_1 and I_2. τ_1 is reduced from the bulk lifetime by an amount that depends on the trapping rate of defects, while τ_2 represents the defect lifetime. Obtained lifetimes and intensities are summarized in tables 1 and 2. Variances of the fit are also included in the tables. Undoped YAG samples were measured only two times; while measurements on Ce doped YAG were repeated five times to confirm the large value of τ_2. Bulk and average lifetimes were calculated using equations (3) and (4).

$$\tau_b = \left(\frac{I_1}{\tau_1} + \frac{I_2}{\tau_2} \right)^{-1} \tag{3}$$

$$\tau_{avg} = \Sigma_i I_i . \tau_i \tag{4}$$

Table 1 and 2 shows that the defect lifetime τ_2 for Ce doped YAG is about 560 ps, much larger than τ_2 for the undoped material (342 ps). The ratio τ_2 / τ_B can was also estimated. It can give some indication about the nature of open volume defects. It is 1.9 for the undoped ceramic and 3.5 for Ce:YAG ceramic revealing much larger open volume defects in Ce:YAG samples. The reason for that could be the large size of Ce ion, as it exists in Ce^{3+} (its size is 1.1 °A) compared to 0.93 °A for Y^{3+} [18]. This probably leads to the formation of vacancy clusters next to the Ce ion, a phenomena we have noticed earlier in Ce doped YAG single crystals [18, 19]. The large defect lifetime in Ce:YAG ceramics also indicates the presence of micro voids and large grain boundaries in the structure. This is an indication that the porosity of Ce:YAG ceramics is higher compared to undoped YAG ceramics, which is consistent with the optical transmission data described below. In case of CeYAG, sintering process probably did not eliminate all the pores at the grain boundaries.

Table 1. PLT data for five runs on Ce:YAG transparent ceramics. τ_1 represents the reduced PLT in bulk material with respective annihilation fraction I_1, τ_2 represents the positron lifetime of defects with respective annihilation fraction I_2, τ_b is the calculated bulk lifetime and τ_{avg} is the average lifetime.

Ce:YAG	**τ_1 (ps)**	**τ_2 (ps)**	**I_1**	**I_2**	**τ_B (ps)**	**τ_2/τ_{Bulk}**	**τ_{avg} (ps)**	**Fit Var**
Run 1	154.9 (0.2)	664.8 (11.2)	97.99% (0.058)	2.01% (0.058)	157.3	4.22632	165.1	1.057
Run 2	154 (0.2)	612.7 (6.4)	97.45% (0.048)	2.55% (.048)	157	3.90255	165.7	1.015
Run 3	153.1 (0.8)	466.1 (9.5)	91.53% (0.377)	8.47% (0.377)	162.3	2.87184	179.6	1.042
Run 4	155.3 (0.4)	532.4 (8)	94% (0.18)	6% (0.18)	162.2	3.28237	177.9	1.092
Run 5	157.9 (0.3)	522.3 (5.3)	93.16% (0.14)	06.84% (0.14)	165.8	3.15018	182.8	1.089
Averages	155	559.7	94.83%	5.17%	160.9	3.48665	174.2	

Table 2. PLT data for two runs on YAG transparent ceramics. τ_1 represents the reduced PLT in bulk material with respective annihilation fraction I_1, τ_2 represents the positron lifetime of defects with annihilation fraction I_2, τ_b is the calculated bulk lifetime and τ_{avg} is the average lifetime.

YAG	**τ_1 (ps)**	**τ_2 (ps)**	**I_1**	**I_2**	**τ_B**	**τ_2/τ_B**	**τ_{avg} (ps)**	**Fit Var**
Run 1	141.8 (0.7)	267.6 (1.2)	61.99% (.63)	38.01% (.63)	172.7	1.54951	189.6	1.075
Run 2	172.9 (0.9)	417.7 (40.9)	96.01% (.72)	3.51% (.66)	177.4	2.35457	180.6	1.063
Averages	157.4	342.7	79%	20.76%	175.1	1.95204	185.1	

Optical transmission spectra of undoped and Ce doped YAG transparent ceramics are shown in Figure 2. More than 80% transmission in the region from 200 to 1100 nm can be seen in the spectra. The dips in CeYAG spectra are due to 5d to 4f transitions in Ce^{3+}. The graph also displays the optical transmission spectra for similar single crystals for comparison. It is interesting to observe that YAG ceramics has higher transparency than single crystals. However this is not the case for CeYAG samples. This is also in agreement with PLT data which revealed larger open volume defects and micro voids in CeYAG structure that are expected to increase light scattering.

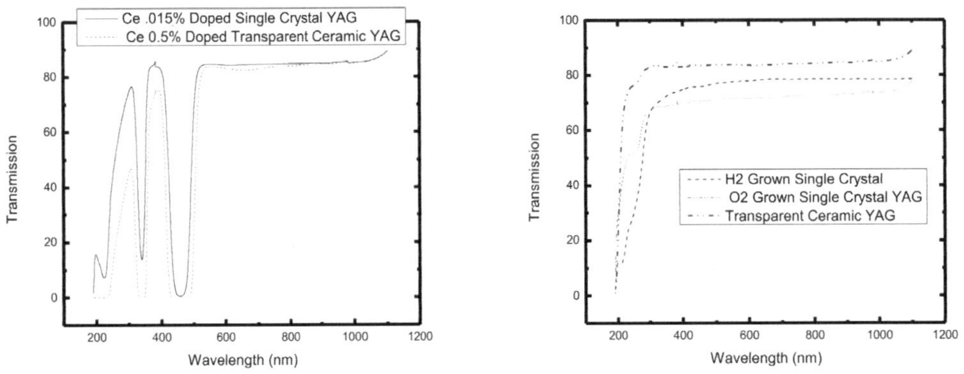

Figure 2. Comparison of optical transmission between transparent ceramic and single crystal YAG and CeYAG samples

4. Conclusions

PLT measurements indicated the absence of large voids and nanopores in undoped and Ce doped YAG transparent ceramics. This explains the reason behind their good transparency (more than 80% transparency in the visible region). However measurements also revealed small open volume defects in the structure, which are much larger for Ce doped samples. Positron annihilation spectroscopy is shown to be a good tool to investigate the microstructures of transparent ceramics and provide a feedback to guide ceramics technology.

5. Acknowledgement

We would like to thank Q. Yuminami at Baikowski / Konoshima International Corp for preparation of YAG and CeYAG transparent ceramic samples. We acknowledge receiving funds from the National Science Foundation under DMR 1359523 grant (Charles Ying) and Bowling Green State University, CURS award (Mora Cordula).

References

[1] Hull G, Roberts JJ, Kuntz J D, Fisher S E, Sanner R D, Tillotson TM , A.D. Drobshoff, Payne S A, Cherepy N J 2007 *Proc. SPIE* **6706** 670617
[2] Wang S F, Zhang J, Luo D W, Gua F, Tang D Y, Dong Z L, Tan G E B, Que W X, Zhang T S, Li S, Kong L B 2013 *Prog. Solid State Chem.* **41** 20
[3] Prochazka I, Cizek J, Melikhova O, Konstantinova T E, Danilenko I A, Yashchishyn I A 2014 *J. Am. Ceram. Soc.* **97** 982
[4] Klym H, Ingram A, Shpotyuk O, Filipecki J Hadzaman I 2011 *J. Phys.: Conf. Ser.* ***289*** *012010 doi:10.1088/1742-6596/289/1/012010*
[5] Fu P, Xu Y, Shi H, Zhang B, Ruan X, Lu W 2014 *Opt. Mater.* **36** 1232
[6] Rempel A, Schaefer H E 1995 *Appl. Phys.* A **61** 51
[7] Zhi Y and Chen A 1993 *J. Phys.: Condens. Matter* **5** 1877
[8] Klym H 2011 *Semicond. Phys. Quantum Electron. Optoelectron.* **14 N 1. P** 109
[9] Drogowska K, Elsayed M, Krause-Rehberg R, Balogh A G 2014 *Ceram. Int.* **40** 9127
[10] Dannefaer S, Bretagnon T, Kerr D 1993 *J. Appl. Phys.* **74** 884
[11] Morsy R, Elsayed M, Krause-Rehberg R, Dlubek G, Elnimr T 2010 *J. Eur. Ceram. Soc.* **30** 1897

[12] Nie J, Xia F, Yu R, Wang B, Li Z, Chen G 2011 *Measurement* **44** 298
[13] Hugenschmidt C, Holzwarth U, Jansen M, Kohn S, Maier K 1997 *J. non-cryst. solids* **217** 72
[14] Krause-Rehberg R, Leipner H: Positron Annihilation in Semiconductors, Springer-Verlag, 1999.
[15] Hautojaervi P: Positrons in Solids, Springer-Verlag, Heidelberg, 1979.
[16] Selim F A, Varney C R, Tarun M C, Rowe M C, Collins G S, McCluskey M D 2014 *Phys. Rev.* B **88** 174102
[17] Winarski D, Persson C, Selim F 2014 *Appl. phys. Lett.* **105** 221110.
[18] Varney C R, Selim FA 2014 *Acta Phys. Pol.* A **125** 764
[19] Selim F A, Winarski D, Varney C, Tarun M C, Ji J, McCluskey D 2015 *Results in Physics* **5** 28

Journal of Physics: Conference Series **674** (2016) 012014 doi:10.1088/1742-6596/674/1/012014

Characterisation of irradiation-induced defects in ZnO single crystals

I Prochazka[1] , J Cizek[1], F Lukac[1], O Melikhova[1], J Valenta[1], V Havranek[2], W Anwand[3], V A Skuratov[4] and T S Strukova[4]

[1] Faculty of Mathematics and Physics, Charles University in Prague, V Holesovickach 2, CZ-180 00 Praha 8, Czech Republic
[2] Nuclear Physics Institute, Academy of Sciences of the Czech Republic, CZ-250 68 Husinec-Rez, Czech Republic
[3] Institute of Radiation Physics, Helmholtz-Zentrum Dresden-Rossendorf, Bautzner Landstr. 400, 01328 Dresden, Germany
[4] Flerov Laboratory of Nuclear Reactions, Joint Institute for Nuclear Research, 141 980 Dubna, Moscow region, Russian Federation

E-mail: ivan.prochazka@mff.cuni.cz

Abstract. Positron annihilation spectroscopy (PAS) combined with optical methods was employed for characterisation of defects in the hydrothermally grown ZnO single crystals irradiated by 167 MeV Xe^{26+} ions to fluences ranged from 3×10^{12} to 1×10^{14} cm^{-2}. The positron lifetime (LT), Doppler broadening as well as slow-positron implantation spectroscopy (SPIS) techniques were involved. The *ab-initio* theoretical calculations were utilised for interpretation of LT results. The optical transmission and photoluminescence measurements were conducted, too. The virgin ZnO crystal exhibited a single component LT spectrum with a lifetime of 182 ps which is attributed to saturated positron trapping in Zn vacancies associated with hydrogen atoms unintentionally introduced into the crystal during the crystal growth. The Xe ion irradiated ZnO crystals have shown an additional component with a longer lifetime of $\approx$ 360 ps which comes from irradiation-induced larger defects equivalent in size to clusters of $\approx$ 10 to 12 vacancies. The concentrations of these clusters were estimated on the basis of combined LT and SPIS data. The PAS data were correlated with irradiation induced changes seen in the optical spectroscopy experiments.

1. Introduction

Zinc monoxide is a direct wide band gap semiconductor predestined for numerous applications: optoelectronic devices, transparent electrodes for solar cells, UV light emitting diodes and lasers, gas sensors *etc*. Application potential of ZnO origins from its large electronic band gap value of $E_g^{(el)} = 3.37$ eV as well as a large exciton binding energy of $E_x \approx 60$ meV at room temperature [1]. Moreover, high-quality bulk ZnO single crystals grown using the hydrothermal (HTG) or pressurised melt growth (MG) techniques have become readily available on the market. Progress in manufacturing the ZnO nanostructures and thin layers has further extended application area of zinc oxide [2].

It is well recognised [1] that functional properties of ZnO based materials, in particular, their optical and electrical features can be strongly influenced by impurity atoms and lattice defects. In this respect, the investigations of structure modifications induced in ZnO by ion implantations deserve attention. There are two main motivations for such investigations: (i) Material properties desired for a particular application may be tailored in this way. (ii) Irradiation-induced degradation of the materials purposed for use in enhanced radiation conditions, *e.g.* in space applications, can be examined. Previous studies of irradiation effects on ZnO were concentrated mainly on the light-particle projectiles [3,4] or implantation of the light ions of energies ranging up to several MeV [5,6]. Information on structure changes induced in ZnO by irradiations with the energetic (≈ 100 MeV) heavy ions is yet rather scarce. To authors' knowledge, only the investigations of zinc oxide bulk crystals irradiated with 100 MeV oxygen and carbon ions [7] and ZnO nanostructured layers irradiated with 130 MeV Xe^{23+} ions [8] have been reported so far. Scanning electron microscopy and optical methods were utilised in these two investigations. Among native impurities in ZnO, hydrogen has to be paid an attention in HTG crystals since H atoms can be easily incorporated into ZnO lattice [9] forming there a shallow donor state [10]. A complex of a zinc vacancy (V_{Zn}) with a H atom, denoted below as V_{Zn}–1H, has been clearly identified as a grown in defect in the HTG bulk ZnO crystals [9,11].

In the present Contribution, positron annihilation spectroscopy (PAS) was used to investigate lattice defects created in the ZnO bulk single crystals irradiated with energetic Xe ions at different fluences. The goal of this investigation was to characterise the irradiation induced defects on a level of atomic arrangement. The conventional PAS, *i.e.* the technique employing β^+ particles from a radionuclide source directly implanted into the sample, was utilised to measure positron lifetimes (LT) and Doppler broadening (DB) of annihilation radiation in its coincidence mode (CDB). In addition, slow-positron implantation spectroscopy (SPIS) with a monoenergetic positron beam was used for determination of positron diffusion lengths. In order to correlate the PAS results with irradiation induced changes in the optical properties of ZnO, two kinds of optical measurements were also performed: optical transmittance (OT) and photoluminescence (PL). A comparison of the present Xe implantation data with those obtained recently for ZnO crystals irradiated with 2.5 MeV protons was also made.

2. Experimental details

2.1. Zinc oxide samples

In the present work, commercial HTG ZnO single crystals of 10×10×0.5 mm^3 dimensions, purchased from MaTecK GmbH, were investigated. The crystals were O-terminated with the (0001) orientation. All their surfaces were optically polished. First, the PAS as well as optical measurements were carried out on the as-grown (virgin) crystals. Then, one face of crystals (denoted below as the front side) was subjected to irradiations. The irradiation by 2.5 MeV protons was performed to a fluence of 10^{16} cm^{-2} using the Tandetron accelerator at NPI Řež. Other series of ZnO crystals were irradiated with 167 MeV Xe^{26+} ions to fluences of 3×10^{12}, 3×10^{13} and 1×10^{14} cm^{-2}. The Xe ion irradiations were made at IC-100 FLNR cyclotron in JINR Dubna. Target temperatures during the proton and Xe irradiations, respectively, did not exceed 100 and 50 °C.

2.2. Conventional PAS methods

Involving the conventional PAS techniques in the investigation of Xe irradiated crystals needs in some substantiation. To get an idea of a depth damaged by the irradiation with Xe ions and protons, the depth profiles of implanted Xe ions, protons and irradiation induced vacancies in ZnO were simulated using the SRIM code [12]. The results of SRIM simulations for the 167 MeV Xe ions and 2.5 MeV protons were plotted against the penetration depth in the ZnO single crystal in figure 1. The stopping profile of ^{22}Na β^+ particles in ZnO, characterised by the mean penetration depth of $z_0 \approx 48$ μm, is also given in the figure for comparison. It is suggested by the data of figure 1 that irradiation damaged regions edge at depths of ≈ 40 and 12 μm, respectively, for protons and Xe ions. Considerable portions of $q \approx 57$ and 21 % of ^{22}Na positrons, respectively, are stopped within damaged regions making thus the conventional PAS to be efficient method in both cases of 2.5 MeV proton and 167 MeV Xe ion irradiations of bulk ZnO single crystals. On the other hand, the conventional PAS can provide only

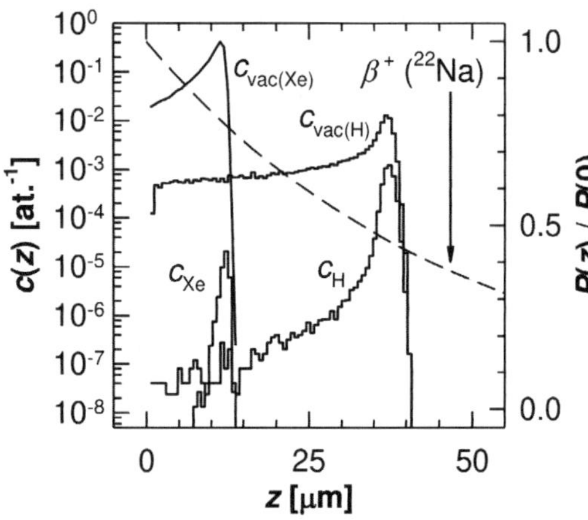

Figure 1. Solid line histograms (the left vertical scale) – the SRIM simulated depth profiles plotted against depth z in ZnO single crystal: 2.5 MeV protons (c_H), 167 MeV Xe ions (c_{Xe}) and vacancies ($c_{vac(H)}$ and $c_{vac(Xe)}$, respectively) created by proton and Xe irradiations.
Dashed line (the right vertical scale) – the stopping profile $P(z)$ of ^{22}Na positrons in ZnO single crystal normalised to its surface value at $z = 0$.

information integrated over a thickness of damaged layer without any sensitivity to the damage depth profile.

Positron sources for the conventional LT and CDB measurements consisted of ≈ 1 MBq of carrier-free ^{22}Na carbonate activity (iThembaLABS production) sealed between 2 μm Mylar® C (Dupont) bearing and covering foils. The source was tightly sandwiched between two identical pieces of samples. In selected cases, both of the front, *i.e.* irradiated, and the rear, *i.e.* unirradiated sides of samples were probed by positrons. The conventional PAS measurements were conducted in air at room temperature.

LT spectra were measured by means of a digital delayed-coincidence spectrometer [13] equipped with BaF_2 scintillators. The spectrometer exhibited a time resolution of 145 ps (FWHM for ^{22}Na energy windows). At least 10^7 coincidence events were collected in each LT spectrum. Measured LT spectra were decomposed into discrete exponential components using a code [14] maximizing the likelihood function. Contribution of annihilations from positron source was derived on the basis of reference LT measurements with a well-annealed α-Fe sample with intensity recalculation for ZnO according to a procedure suggested in [15].

CDB spectra were acquired using a digital HPGe–HPGe coincidence spectrometer [16] exhibiting an energy resolution of 0.9 keV (FWHM in coincidence mode at 511 keV γ-ray energy). Each two-dimensional (2D) energy spectrum contained at least 10^8 coincidence events. The one-dimensional Doppler-broadened profiles (DBPs) were then extracted from the 2D spectra and displayed as the ratios to the DBP shown by a well-annealed pure Al specimen. A peak-to-background ratio observed for DBPs was better than 10^5. In addition, the CDB spectra for the well-annealed high-purity zinc metal and alumina (Al_2O_3) were taken for sake of comparisons.

2.3. SPIS measurements

SPIS measurements were performed using a magnetically guided variable-energy slow-positron beam SPONSOR [17]. The energies of incident monoenergetic positrons covered a range from 30 eV to 35 keV corresponding to a mean penetration depth of positrons in ZnO ranging from the surface up to 2.3 μm. The DB shape of the 511 keV annihilation peak was measured with a HPGe γ-spectrometer exhibiting an energy resolution of 1.09 keV (FWHM) at this energy. Typically, $\approx 5\times10^5$ counts were collected in the 511 keV peak. The DB peak shapes were characterised by the ordinary sharpness (S) and wing (W) parameters [18] evaluated as functions of positron energy E_+.

2.4. Optical methods

An UV-VIS spectrometer (Ocean Optics) equipped with a deuterium tungsten halogen light source and a grating monochromator was employed in OT measurements. The measurements were conducted at room temperature within a wavelength interval from 340 through 880 nm. To a first approximation, the reflectivity of each crystal was considered to be independent of wavelength so that no corrections of the transmitted light intensity for reflectivity variations with wavelength were applied.

PL apparatus was composed of an Olympus IX-71 inverted microscope with an Ealing 25×/0.4 mirror objective and an ARC SP-2300i imaging spectrometer. Samples were irradiated by an UV LED lamp generating 325 or 310 nm photons. A LN-cooled back-illuminated CCD was used to detect luminescence. A lateral resolution of ≈ 0.5 μm was exhibited by the imaging system. The PL spectra were measured at room temperature.

3. Results and discussion

In table 1, positron lifetimes τ_i and relative intensities I_i of exponential components, recognised in the measured LT spectra as arising from positron annihilations in ZnO samples, are shown (i = 2,3,4). Not shown were two weak components that originated from annihilations in the positron source exhibiting lifetimes (intensities) 368 ps (9 %) and 1.8 ns (2 %).

Table 1. Positron lifetimes τ_i and relative intensities I_i (i=2,3,4) of exponential components resolved in the LT spectra of ZnO. Fluences f are shown in the second column, in which symbols 'r' indicate measurements of the rear side of irradiated samples. The intensities are normalised so that $\sum_i I_i = 100$ %. The standard deviations are given in parentheses in the units of the last significant digit.

ZnO sample	f [cm^{-2}]	τ_2 [ps]	I_2 [%]	τ_3 [ps]	I_3 [%]	τ_4 [ps]	I_4 [%]
virgin		183.2 (1)	100				
Xe^{26+} irrad.	3×10^{12}	184 (1)	86.3 (8)			369.6(6)	13.7(7)
	3×10^{13}	185.4 (5)	78.9 (4)			365(2)	21.1(5)
	1×10^{14}	184.1 (7)	74.5 (8)			351(1)	25.5(7)
	r 1×10^{14}	183 (2)	100				
H^+ irrad.	1×10^{16}	175 (5)	71 (2)	260 (2)	29 (2)		
	r 1×10^{16}	180 (1)	100				

In all the samples of table 1, the shortest observed lifetime τ_2 appeared to exhibit a value around 182 ps. In the virgin ZnO as well as in both irradiated samples faced to the positron source by their rear side, τ_2-component was the only component arising from annihilations in ZnO crystal. When the irradiated ZnO samples were faced to the source by their front side, an additional component with lifetime $\tau_3 \approx 260$ ps and $\tau_4 \approx 360$ ps, respectively, appeared for the proton and Xe ion implanted samples, representing thus a purely irradiation-induced effect.

There was a long-lasting debate about lifetime τ_B exhibited by delocalised positrons in the perfect zinc oxide lattice. Several authors [5,18] claimed that $\tau_B \approx 150$ ps while some others [3] stated τ_B to be as high as ≈ 180 ps. Comprehensive treatment of these discrepancies, taking into account all available experimental as well as theoretical data, was undertaken by Brauer et al. [9]. Based on their argumentation, we adopt the lifetime of delocalised positrons in the perfect ZnO lattice to be $\tau_B \approx 152$ ps. Since positron lifetimes given in table 1 significantly exceed this τ_B-value, their origin has fully to come from saturated positron trapping in defects, *i.e.* no components

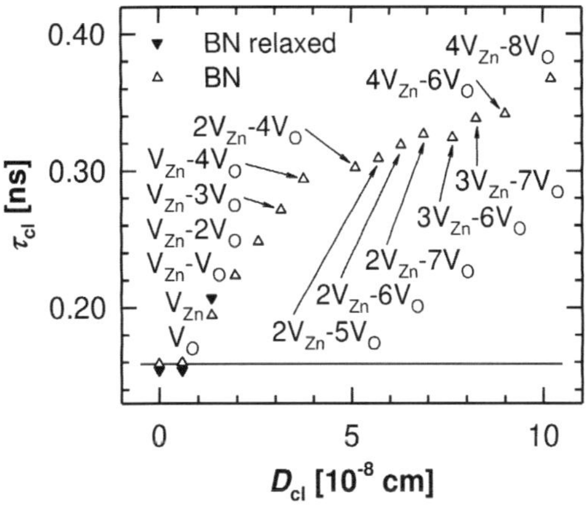

Figure 2. Calculated positron lifetimes τ_{cl} for selected vacancy clusters in ZnO plotted against cluster size D_{cl} (see the text for further details).

Journal of Physics: Conference Series **674** (2016) 012014 doi:10.1088/1742-6596/674/1/012014

corresponding to delocalised positrons (τ_1) could be observed. The present τ_2-components undoubtedly arise from annihilations of positrons trapped in the grown-in V_{Zn}–1H complexes which were clearly proven earlier to exist in the HTG ZnO crystals [9,11]. The τ_3-value observed in the proton irradiated sample, see table 1, appears to be close to the lifetime ≈ 260 ps observed in [5] in the 10 MeV electron irradiated ZnO and attributed there to positron trapping in the irradiation-induced V_{Zn}–V_O divacancies. Because damage induced by 2.5 MeV protons is expected to be similar to that produced by 10 MeV electrons, it is natural to assign the present proton-induced τ_3-component to positrons trapped in V_{Zn}–V_O divacancies, too.

The τ_4-lifetime of ≈ 360 ps resulting from Xe^{26+} irradiations (see table 1) obviously belongs to positron trapping in larger open-volume defects. To estimate their size, a variety of selected agglomerates of vacancy-like defects, composed of m zinc vacancies and n oxygen vacancies, mV_{Zn}–nV_O, was modelled and respective positron lifetimes were calculated by *ab-initio* calculations using the atomic superposition (ATSUP) method [19]. The electron-positron correlation has been included within the Boroński-Nieminen (BN) approach with the correction for incomplete positron screening taking the high-frequency dielectric constant as $\varepsilon_\infty = 4$ [20]. In figure 2, the positron lifetimes calculated for various vacancy-like agglomerates mV_{Zn}–nV_O were plotted against agglomerate size parameter D_{cl} introduced as a diameter of a sphere which volume is equivalent to a sum of volumes of m V_{Zn} and n V_O monovacancies. The data of figure 2 suggest that observed τ_4-lifetimes best correspond to vacancy agglomerates of open volume equivalent to ≈ 10 to 12 monovacancies.

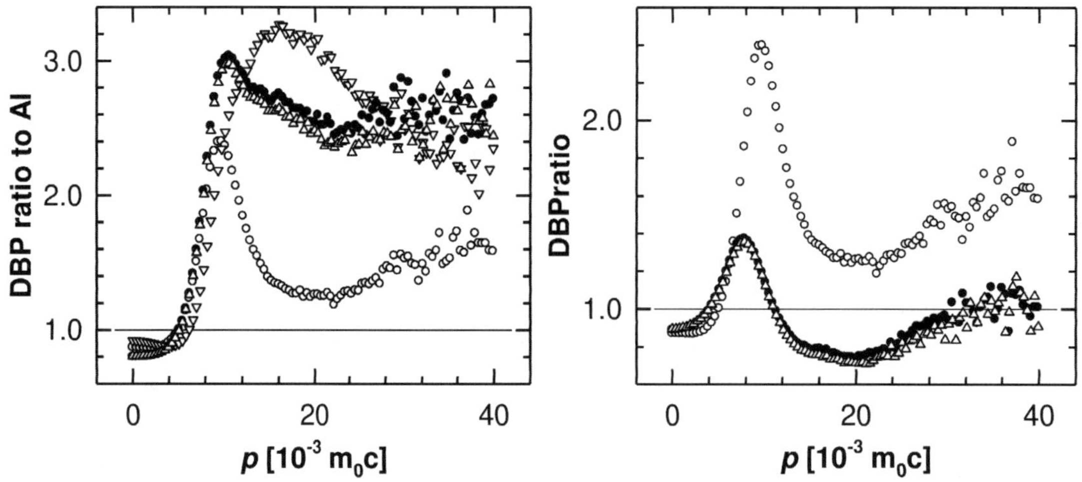

Figure 3. DBP ratios plotted as functions of electron momentum p : , • – virgin ZnO crystal, Δ – Xe irradiated ZnO crystal (1×10^{14} cm^{-2}), ○ – Al_2O_3 , ∇ – pure zinc metal. The left panel: All the DBPs are related to well-annealed pure Al metal. The right panel: Al_2O_3 is related to well-annealed pure Al metal (as in the left panel), but virgin as well as irradiated ZnO are related to well-annealed pure Zn metal.

The present CDB results are displayed as the DBP ratios to a well-annealed pure Al or Zn metals in figure 3 where they are compared to the DBP ratios measured for the high purity zinc metal and alumina (Al_2O_3). Note that DBP ratio for Al_2O_3 with respect to Al is expected to exhibit merely a contribution which is characteristic of 2p oxygen electrons in a similar way as the ratio curve for SiO2 with respect to Si [21]. The figure thus suggests that a dominating effect observed in the CDB spectra of ZnO samples comes from oxygen 2p electrons.

The SPIS results of the present work are shown in figure 4. The measured S-parameters of the figure were fitted with the VEPFIT code [22] giving the positron diffusion lengths collected in table 2. Combination of the SPIS and LT results can be utilised to estimate the concentrations of irradiation

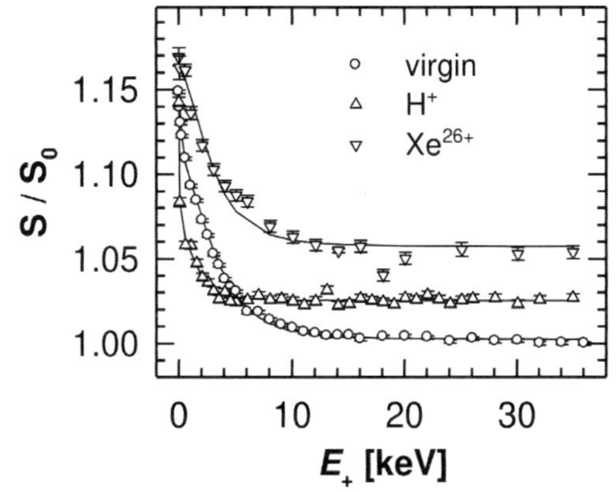

Figure 4. Measured *S*-parameters as functions of positron energy E_+. Solid lines represent VEPFIT calculations. The *S*-values are normalised to the fitted bulk *S*-parameter for virgin ZnO, $S_0 = 0.5004$.

Table 2. VEPFIT evaluations of SPIS data: L_+ – positron diffusion length, S_{bulk}/S_0 – bulk *S*-parameter normalized to that of virgin ZnO.

sample	f [cm^{-2}]	L_+ [nm]	S_{bulk}/S_0
virgin		58 ± 2	1.0000
Xe^{26+} irrad.	1×10^{14}	47 ± 2	1.0574±0.0007
H^+ irrad.	1×10^{16}	19 ± 2	1.0253±0.0007

induced defects. First, the concentration $c_{V_{Zn}-1H}$ of the V_{Zn}–1H defect in the virgin ZnO can be estimated from the measured positron diffusion length L_+ of table 2 using the expression [22]

$$c_{V_{Zn}-1H} = \frac{1}{\mu_{V_{Zn}-1H}\,\tau_B}\left(\frac{L_{+,B}^2}{L_+^2}-1\right), \qquad (1)$$

where $L_{+,B}$ and τ_B denote the positron diffusion length and lifetime in a perfect zinc oxide lattice, respectively, and $\mu_{V_{Zn}-1H}$ is the positron specific trapping rate for the V_{Zn-1H} defect. Inserting into eq. 1 $\tau_B \approx 152$ ps [9] and the values typical for semiconductors [23], viz.

$L_{+,B} \approx 200$ nm and $\mu_{V_{Zn}-1H} \approx 1\times10^{15}$ s^{-1},

the $c_{V_{Zn}-1H}$ is estimated as $\approx 7\times10^{-5}$ at.$^{-1}$. Then the concentration c_{cl} of the Xe irradiation induced clusters can be estimated within the simple trapping model [24] as

$$c_{cl} = \frac{1}{q}\frac{I_4}{I_2}\frac{\Omega_{V_{Zn}-1H}}{\Omega_{cl}}\mu_{V_{Zn}-1H}\,c_{V_{Zn}-1H}\,. \qquad (2)$$

In this equation, $\Omega_{V_{Zn}-1H}$ and Ω_{cl} stand for volumes of the V_{Zn-1H} and cluster, respectively, $\Omega_{cl}/\Omega_{V_{Zn}-1H} \approx \Omega_{4V_{Zn}-8V_O}/\Omega_{V_{Zn}-1H} \approx 4.7$. Factor q accounts for a fraction of positrons that are stopped within the depth damaged by Xe ions. The concentration c_{cl} of Xe ions induced clusters, estimated using eq. 2 with $q \approx 0.2$ indicated by the above SRIM simulations, was plotted in figure 5 against the Xe ion fluence. One can notice in the figure a trend to a plateau-like pattern displayed by c_{cl}-values when the fluence increases to $\approx 1\times10^{14}$ cm^{-2}. An overlapping of Xe ion tracks may lie behind the cluster creation.

Measured transmittances T are represented as functions of wavelength λ in figure 6. For the virgin ZnO crystal, T-curve exhibits a shape which is essentially the same as that reported earlier for HTG ZnO single crystals [25]. A rapid onset of T occurs at $\lambda \approx 390$ nm with T risen up to $T \approx 60$ % at $\lambda \approx 410$ nm. Then the slope of T is gradually dropped down and finally a very slow and almost linear increase of T with increasing λ is observed above $\lambda \approx 550$ nm, reaching a maximum value of $T \approx 80$ % at $\lambda \approx 880$ nm.

For the proton-implanted ZnO crystal, the observed transmittance closely follows that found for the virgin crystal, except for the 400 to 550 nm wavelength range (figure 6). At $\lambda \approx 400$ nm ($T \approx 25$ %), the observed transmittance in the proton-implanted ZnO becomes significantly reduced with respect to the virgin crystal, growing roughly linearly up to a value of $T \approx 75$ % at $\lambda \approx 550$ nm, i.e.

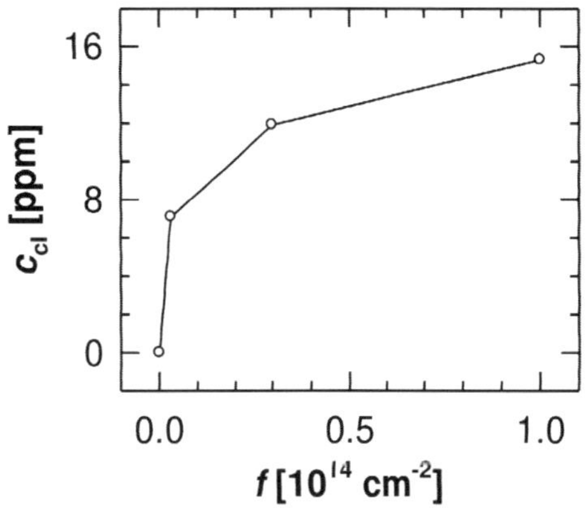

Figure 5. Estimated concentrations c_{cl} of Xe ion irradiation induced clusters in ZnO vs Xe ion fluence f.

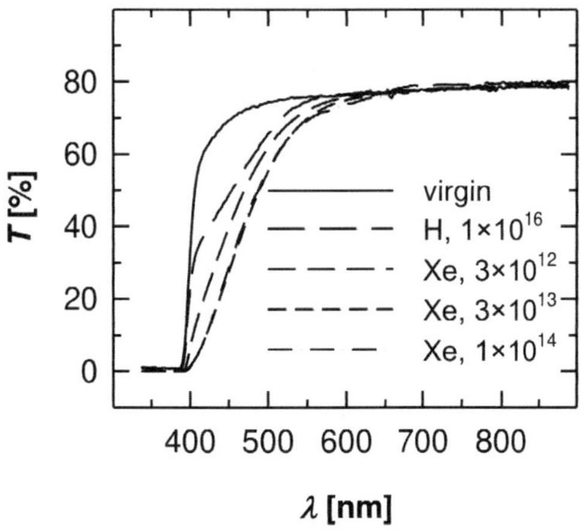

Figure 6. Measured transmittances T as functions of wavelength λ.

reaching again a T-values found in the virgin ZnO (see figure 6). Thus defects which are optically active in the violet – blue – green spectral region are brought into the ZnO crystals by 2.5 MeV proton irradiation. Protons of such relatively low energy are capable to produce only atomic size defects (V_{Zn} or V_O, interstitials, divacancies) in the subsurface layer of ≈ 40 μm depth of ZnO crystals. It is obvious that some of these point defect types should be regarded as a cause of the observed suppression of transmittance within a 410 to 550 nm region. Of them, however, only V_{Zn} and V_{Zn}–V_O defects could be detected by PAS and V_{Zn}–V_O defects were indeed observed in proton irradiated ZnO crystal, see table 1.

In the Xe-implanted ZnO crystals, irradiation-induced suppression of transmittance inside a 410 to 550 nm region is even more pronounced than in the proton-implanted sample. Transmittance is further reduced for a fluence of $f \approx 3\times10^{13}$ cm^{-2} compared to that of $f \approx 3\times10^{12}$ cm^{-2}, see figure 6, showing on the other hand a plateau effect at a fluence of ≈ 1×10^{14} cm^{-2}. Note that an approaching to a plateau is also seen in cluster concentration data discussed in the preceding text. Vacancy clusters created inside a 12 μm subsurface layer by Xe ion irradiation of ZnO crystals should be considered as a likely cause of observed transmittance dropping down in this case.

The PL spectra taken for the virgin and implanted ZnO crystals are displayed as functions of wavelength λ in figure 7. A green emission (GE) band with a broad maximum at $\lambda \approx 550$ nm [1] is clearly seen for the virgin ZnO crystal. For both irradiation species, however, the GE becomes completely suppressed to a level of scattering background of the spectrometer. Despite of a great volume of research work, the nature of GE band in ZnO remains still not fully understood [26]. Nonetheless, the present PL experiments have suggested that the observed GE quenching is likely a consequence of irradiation induced defects of sufficient concentrations providing a channel for dominating non-radiative recombination of charge carriers.

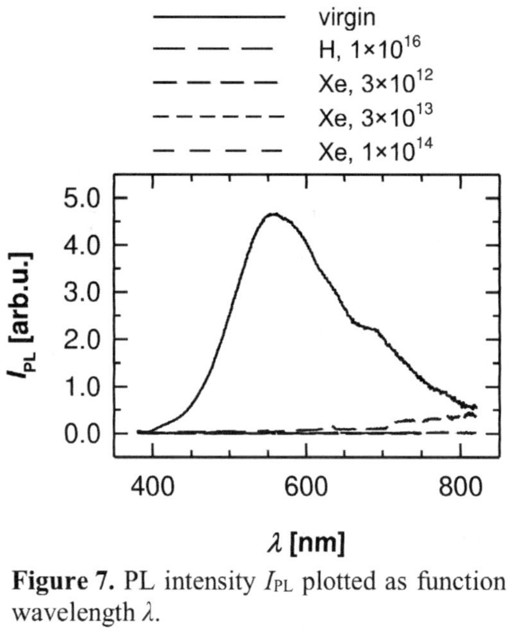

Figure 7. PL intensity I_{PL} plotted as function of wavelength λ.

4. Conclusions

In the present work, hydrothermally grown bulk ZnO single crystals were implanted with the 2.5 MeV protons to a fluence of 1×10^{16} cm^{-2} and the 167 MeV Xe^{26+} ions to fluences of 3×10^{12}, 3×10^{13} and 1×10^{14} cm^{-2}. Irradiation induced changes were studied by means positron annihilation techniques (positron lifetime, coincidence Doppler broadening and slow-positron implantation spectroscopy measurements) combined with optical methods (optical transmittance, photoluminescence). The PAS results were correlated with those obtained by the optical methods.

The grown in as well as irradiation induced defects in ZnO crystals were characterised: the grown in V_{Zn}–1H complexes, the V_{Zn}–V_O divacancies created in the proton irradiated crystals, and the larger defects of size comparable to clusters of ≈ 10 to 12 vacancies identified in the Xe implanted crystals. Concentrations of the observed defects were estimated by combining LT and SPIS results. The irradiation induced defects were demonstrated to be optically active in the 400 to 550 nm wavelength region. The green emission band which is well-seen in the virgin ZnO crystal was completely destroyed in the both proton and Xe ion irradiated samples which may be likely connected with irradiation-induced channels for non-radiation recombination of charge carriers.

Trend to a plateau-like behaviour of irradiation induced changes at a Xe ion fluence of $\approx 1\times10^{14}$ cm^{-2}, exhibited by the present OT data as well as cluster concentration estimates, may be tentatively regarded as an evidence of strongly overlapping heavy ion tracks at sufficiently high fluences when no additional increase of structure disorder in the damaged subsurface layer can obviously be produced by increasing fluence. Similar considerations have recently been reported for heavy ion implanted sapphire crystals [27]. Indeed, further studies of these interesting phenomena may be of a significant theoretical as well as practical interest.

Acknowledgements
Present investigation was financially supported by the Czech Science Foundation under the project P108/11/0958. Also financial support of the cooperation of the Czech Republic with JINR within the '3+3' project is highly acknowledged.

References
[1] Özgür Ü, Alivov Ya I, Liu C, Teke A, Reshchikov M A, Doğan S, Avrutin V, Cho S-J and Morkoç H 2005 *J. Appl. Phys.* **98** 041301
[2] Ursaki V V, Rusu E V, Sarua A, Kuball M, Stratan G I, Burlacu A and Tiginyanu I M 2007 *Nanotechnology* **18** 215705
[3] Tuomisto F, Ranki V, Saarinen K and Look D C 2003 *Phys. Rev. Lett.* **91** 205502
Tuomisto F, Saarinen K, Look D C and Farlow G C 2005 *Phys. Rev.* B **72** 085206
[4] Brunner S, Puff W, Balogh A G and Mascher P 2001 *Mater. Sci. Forum* **363–365** 141
[5] Brauer G, Anwand W, Skorupa W, Kuriplach J, Melikhova O, Moisson C, von Wenckstern H, Schmidt H, Lorenz M and Grundmann M 2006 *Phys. Rev.* B **74** 045208
[6] Ronning C, Gao P X, Ding Y, Wang Z L and Schwen S 2004 *Appl. Phys. Lett.* **84** 783
[7] Regel' V P, Nikitenko V A, Kuzmina I P, Galstyan V G, Dolukhanyan T P, Nikul'shin S F, Sizova N L and Skuratov V A 1987 *Sov. Phys. – Tech. Phys.* **32** 183
[8] Burlacu A, Ursaki V V, Skuratov V A, Lincot D, Pauporte T, Elbelghitt T, Rusu E V and Tiginyanu I M 2008 *Nanotechnology* **19** 215714
[9] Brauer G, Anwand W, Grambole D, Grenzer J, Skorupa W, Čížek J, Kuriplach J, Procházka I, Ling C C, So C K, Schulz D and Klimm D 2009 *Phys. Rev.* B **79** 115201
[10] Van de Walle C G 2001 *Physica B* **308–310** 899
[11] Cizek J, Zaludova N, Vlach M, Danis S, Kuriplach J, Prochazka I, Brauer G, Anwand W, Grambole D, Skorupa W, Gemma R, Kirchheim R and Pundt A 2008 *J. Appl. Phys.* **103** 053508
[12] Ziegler J F, Ziegler M D and Biersack J P 2010 *Nucl. Instr. Meth in Phys. Research* B **268** 1818 Information on www.srim.org
[13] Becvar F, Cizek J, Prochazka I and Janotova J 2005 *Nucl. Instr. Meth. in Phys. Research* A **539** 372
[14] Prochazka I, Novotny I and Becvar F 1997 *Mater. Sci. Forum* **255–257** 772
[15] Surbeck H 1977 *Helv. Phys. Acta* **50** 705
[16] Cizek J, Vlcek M and Prochazka I 2010 *Nucl. Instr. Meth. in Phys. Research* A **623** 982
[17] Anwand W, Brauer G, Butterling M, Kissener H R and Wagner A 2012 *Defects and Diffusion Forum* **331** 25

[18] Mizuno M, Araki H and Shirai Y 2004 *Materials Transactions* **45** 1964
[19] Puska M J and Nieminen R M 1983 *J. Phys.* F **13** 333
Seitsonen A P, Puska M J and Nieminen R M 1995 *Phys. Rev.* B **51** 14057
[20] Boronski E and Nieminen R M 1986 *Phys. Rev.* B **34** 3820
Puska M J, Mäkinen S, Manninen M and Nieminen R M 1989 *Phys. Rev.* B **39** 7666
[21] Brusa R S, Deng W, Karwasz G P, Zecca A and Pliszka D 2001 *Appl. Phys. Lett.* **79** 1492
[22] Van Veen A, Schut H, Clement M, de Nijs J M M, Kruseman A and IJpma M 1995 *Appl. Surf. Sci.* **85** 216
Van Veen A, Schut H, de Vries J, Hakvoort R A and IJpma M R 1990 *AIP Conf. Proc.* **218** 171
[23] Krause-Rehberg R and Leipner H S 1999 *Positron Annihilation in Semiconductors – Defect Studies* (Berlin: Springer)
[24] West R N 1979 *Positrons in Solids* ed P Hautojärvi (Berlin: Springer-Verlag) p 89
[25] Anwand W, Brauer G, Grynszpan R I, Cowan T E, Schulz D, Klimm D, Cizek J, Kuriplach J, Prochazka I, Ling C C, Djurisic A B, Klemm V, Schreiber G and Rafaja D 2011 *J. Appl. Phys.* **109** 063516
[26] Sun Y, Guo H, Jiang F, Yuan R, Zhang J, Zeng X, Zhou T, Qiu Y, Zhang B, Xu K and Yang H, 2013 *Appl. Surf. Sci.* **283** 258
[27] Liszkay L, Gordo P M, Havancsak K, Skuratov V A, de Lima A and Kajcsos Zs 2004 *Mater. Sci. Forum* **445–446** 138

Study on lattice defects in CeO_2 by means of positron annihilation measurements

Y Yamamoto, T Kishino, T Ishiyama, A Iwase, F Hori

Department of Materials Science, Osaka Prefecture University, Sakai, Osaka 599-8531, Japan

E-mail: horif@mtr.osakafu-u.ac.jp

Abstract. Positron annihilation coincidence Doppler broadening (CDB) measurements were performed for CeO_2 after the high temperature annealing or the electron irradiation. Both the annealing and the electron irradiation, the vacancy-type defects were detected by positron annihilation measurement. In contrast, CDB results showed the different type of vacancies was introduced by annealing and electron irradiation.

1. Introduction

CeO_2 has been used as a simulation material of UO_2 nuclear fuels in order to study the effect of energetic fission fragments on the structure of nuclear fuels [1, 2]. There are close similarities between CeO_2 and UO_2 in terms of melting temperatures, lattice constants, and crystal structures [3]-[6]. Recently, we have reported the effects of 200 MeV xenon ion irradiation, 10 MeV iodine ion irradiation and annealing at 1273 K on magnetic property of CeO_2 pellets [7, 8]. After the irradiation or the high temperature annealing, we observed that the ferromagnetic state appeared. We suggested that the appearance of ferromagnetic state in CeO_2 is caused by the formation of lattice defects, especially oxygen vacancies. The structures of defects which are produced by the annealing or the ion irradiations have, however, never been clarified yet.

In this paper, we have performed the positron annihilation coincidence Doppler broadening (CDB) measurement in order to estimate the defect structure in annealed and electron-irradiated CeO_2. This method is to highlight the differing elemental environments around defects in which positrons get trapped and can be highly useful in studies involving defect-related properties such as magnetism, heat treatment, and so on [9]. In the present experiment, the X-ray diffraction (XRD) measurement was also carried out to study the change in lattice structure.

2. Experimental procedure

Specimens were as-cast CeO_2 bulk pellets which were prepared by sintering CeO_2 powder. The size of these specimens was about 15 mm in diameter and 0.7 mm thick. They were annealed at 1273 K for 1 hour in a vacuum of 1.0×10^{-4} Pa. On the other hand, other CeO_2 specimens were cut into two pieces of 3 mm squared. They were irradiated with 8 MeV electrons at room temperature by using the LINAC at Kyoto University Research Reactor Institute. The fluence of electron irradiation was 5.0×10^{17} / cm^2 and 2.0×10^{18} / cm^2 for each. Annealed and electron-irradiated specimens were characterized by using a conventional Cu-Kα X-ray diffraction. These CeO_2 specimens and pure cerium were performed CDB measurements at room temperature in the atmosphere. A ^{22}Na positron source was covered with thin

Journal of Physics: Conference Series **674** (2016) 012015 doi:10.1088/1742-6596/674/1/012015

kapton films. For irradiated samples, we measured CDB spectrum by sandwiching two samples with the fluences of 5.0×10^{17} /cm^2 and 2.0×10^{18} /cm^2 simultaneously, as the average fluence of 1.25×10^{18} /cm^2 irradiation. More than 1.0×10^8 coincidence counts were acquired under each spectrum. In the present analysis, S and W parameter was defined as $|P_L| < 5.0\times10^{-3}\ m_0c$ and $12.5\times10^{-3}\ m_0c < |P_L| < 20.5\times10^{-3}\ m_0c$ in the CDB spectrum, respectively.

3. Results and discussion

Figure 1 shows the photograph of as-cast, annealed and electron-irradiated CeO_2 specimens. The color of CeO_2 changed to black by the annealing. On the other hands, the color of CeO_2 slightly changed to brown by the electron irradiation. There are some reports that the color of oxide materials such as CeO_2 [10, 11] and TiO_2 [12] changed because of vacancy-type of defects. Taking into account for these facts, the annealing and the electron irradiation may introduce vacancy-type of defects in the CeO_2 specimens.

The defect concentration by the electron irradiation with the fluence of 1.25×10^{18} / cm^2 is roughly estimated 10^{-4} dpa. On the other hands, the defect concentration by the annealing [13] has a bigger one or two digit than that by the electron irradiation. The number of vacancies in annealed CeO_2 is much more than that in electron irradiated CeO_2. From this result, the color changes with increasing vacancy-type of defects.

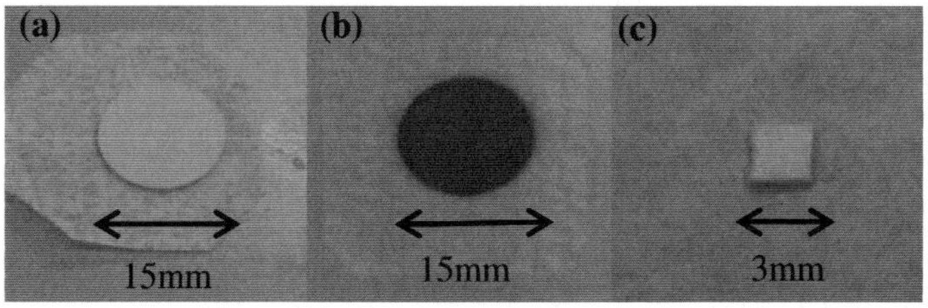

Figure 1. The photograph of (a) as-cast CeO_2, (b) annealed CeO_2, and (c) electron irradiated CeO_2.

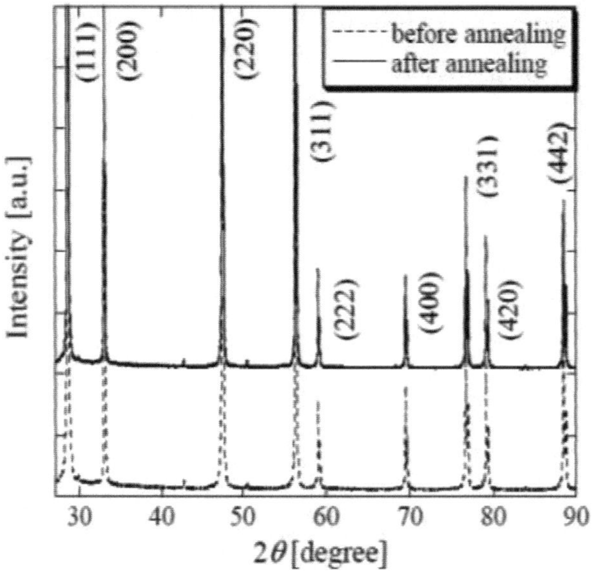

Figure 2. Widely-scanned XRD spectra for CeO_2 before and after annealing.

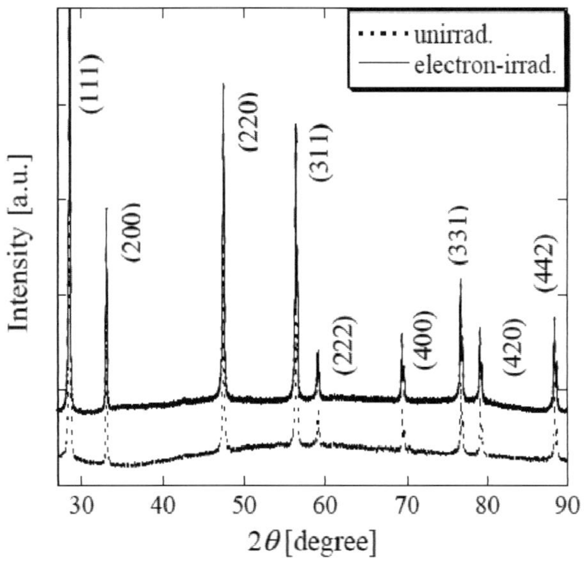

Figure 3. Widely-scanned XRD spectra for unirradiated and electron-irradiated CeO_2 specimens.

Figure 2 shows the XRD profiles for CeO_2 specimens before and after the annealing. XRD peaks appear only the peaks of fluorite structure even after the annealing. Figure 3 shows the XRD profiles for unirradiated CeO_2 specimens and those irradiated with electrons. XRD peaks appear the peaks of fluorite structure even after the electron irradiation. The XRD results show that in spite of the annealing and the electron irradiation, the lattice structure remains the fluorite structure.

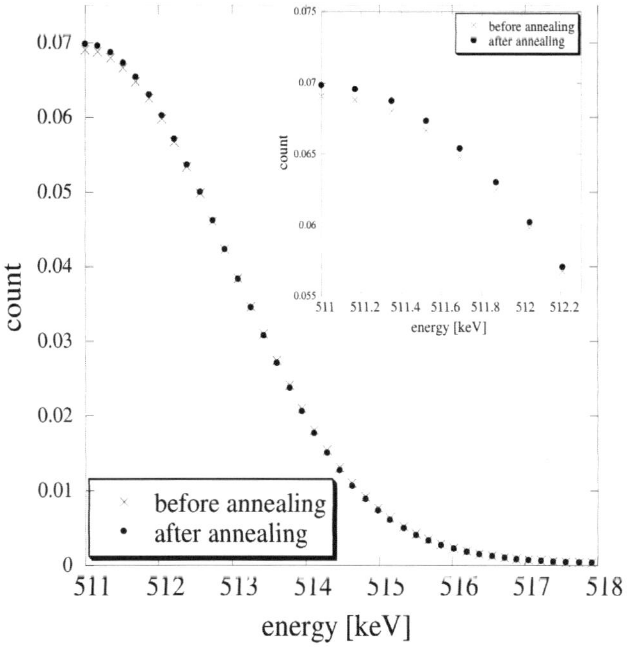

Figure 4. Doppler spectra around 511 keV for CeO_2 before and after annealing. Enlarged graph around 511 keV is also shown.

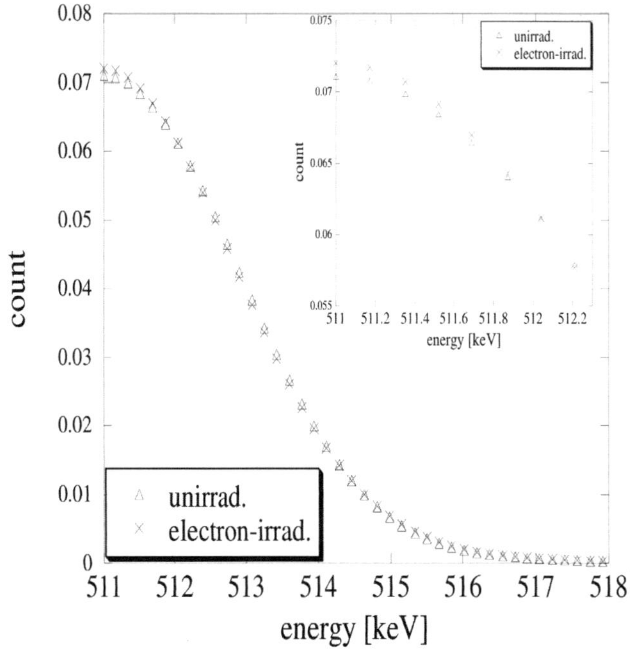

Figure 5. Doppler spectra around 511 keV for CeO_2 before and after electron irradiation. Enlarged graph around 511 keV is also shown.

Figures 4 and 5 show the positron annihilation Doppler spectra for as-cast, annealed and electron-irradiated CeO_2. The total count of each Doppler spectrum for these CeO_2 specimens is normalized. As can be seen in the inset of Figure 4 which shows the enlarged spectra around 511 keV, the intensity of Doppler spectrum for the annealed CeO_2 is greater than that for unannealed CeO_2. Figure 5 shows the same tendency in electron-irradiated CeO_2. This result means that the vacancy-type of defects is introduced with the annealing or the electron irradiation.

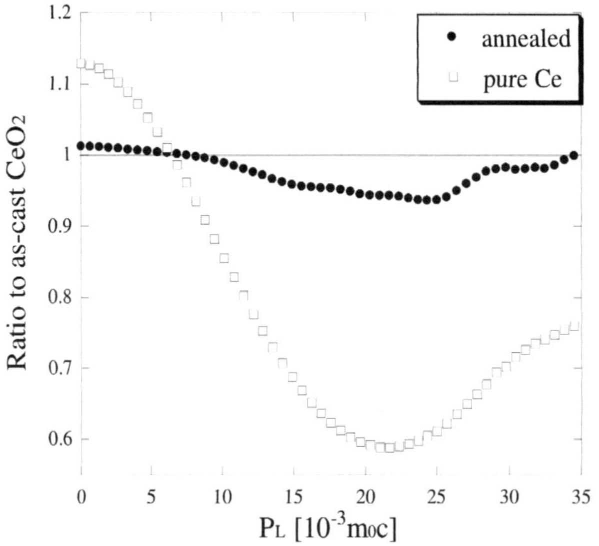

Figure 6. CDB ratio curves related to as-cast CeO_2 for annealed CeO_2 and pure cerium.

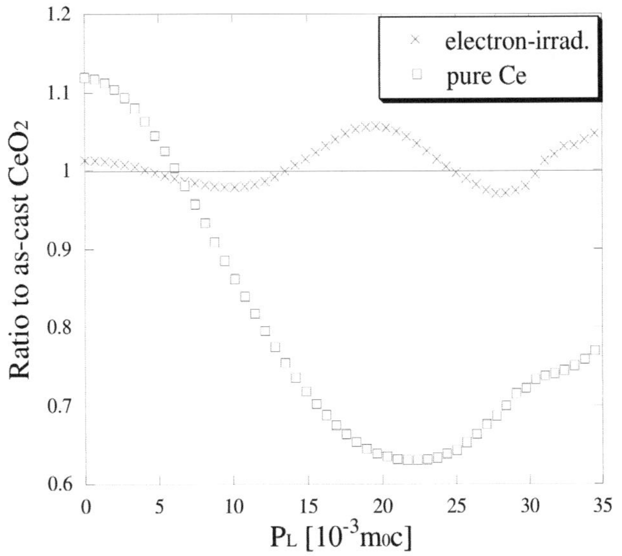

Figure 7. CDB ratio curves related to as-cast CeO_2 for CeO_2 irradiated with 8MeV electron and pure cerium.

Figure 6 shows the CDB ratio curves related to as-cast CeO_2 for annealed CeO_2 and pure cerium. The CDB spectrum for the annealed specimens shows a minimum value around 20×10^{-3} m_0c which is also seen in pure cerium. This result reveals that atoms around the introduced vacancies in the annealed CeO_2 are almost cerium. In other words, oxygen vacancies are mainly produced by the annealing. Figure 7 shows CDB ratio curves related to as-cast CeO_2 for electron-irradiated CeO_2 and pure cerium. In Figure 7, the CDB profile of electron irradiated CeO_2 is quite different from that of pure cerium. This result shows that atoms around the introduced vacancies in electron-irradiated CeO_2 are not necessarily cerium. In other words, the electron irradiation produces not only oxygen vacancies but also cerium vacancies and so on. From this result, atoms around the introduced vacancies are different between the annealing and the electron irradiation.

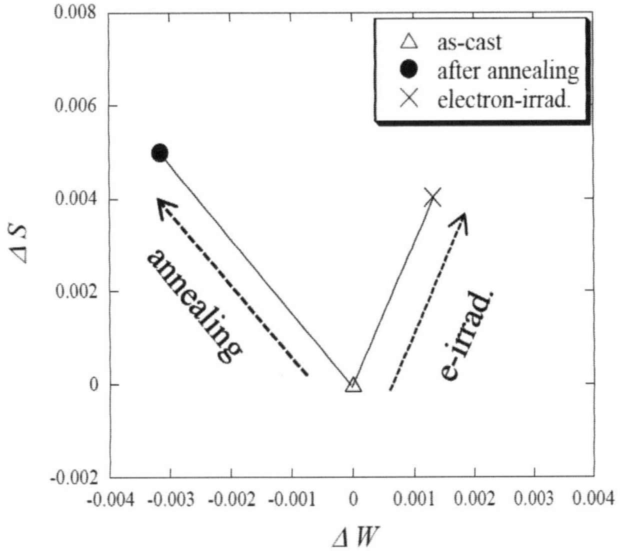

Figure 8. The plots of the change of S-parameter and W-parameter for CeO_2 specimens, as-cast, after annealing, and after electron irradiation.

Figure 8 shows the plots of the change of S-parameter, ΔS, and the change of W-parameter, ΔW, for annealed and electron-irradiated CeO_2 specimens. By the annealing, S parameter increases with decreasing W parameter. On the contrary, S parameter increases with increasing W parameter by the electron irradiation. This result suggests that the different vacancy-type of defects were produced by the annealing or the electron irradiation. The results of CDB ratio curves and S-W plots show that oxygen vacancies are dominantly produced by the annealing, but not only oxygen vacancies but also cerium vacancies produced by the electron irradiation can be detected by positron annihilation measurement.

4. Summary

CeO_2 specimens were annealed in a vacuum or irradiated with 8MeV electrons. The positron annihilation CDB measurements were performed in order to study lattice defects. The CDB results show that both the annealing and the electron irradiation produce the vacancy-type defects. However, the type of vacancies is different between the annealing and the electron irradiation. By the annealing, oxygen vacancies are mainly introduced. On the contrary, not only oxygen vacancies but also cerium vacancies are introduced by the electron irradiation.

Acknowledgements
The authors are grateful to Prof. K. Yasuda of the Departments of Applied Quantum Physics and Nuclear Engineering, Kyushu University for his help.

References

[1] Sonoda T, Kinoshita M, Chimi Y, Ishikawa N, Sataka M, and Iwase A 2006 *Nucl. Instrum. Methods Phys. Res., Sect.* B **250** 254
[2] Sonoda T, Kinoshita M, Ishikawa N, Sataka M, Chimi Y, Okubo N, Iwase A, and Yasunaga K 2008 *Nucl. Instrum. Methods Phys. Res., Sect.* B **266** 2882
[3] Ohhara K, Ishikawa N, Sakai S, Matsumoto Y, Michikami O, and Ohta Y 2014 *Nucl. Instrum. Methods Phys. Res., Sect.* B **267**, 973
[4] Lyons M F, Boyle R F, Davies J H, Hazel V E, and Rowland T C 1992 *Nucl. Eng. Des.* **21** 167
[5] Kang K H, Ryu H J, Song K C, and Yang M S 2002 *J. Nucl. Mater.* **301** 242
[6] Schwab R G, Steiner R A, Mages G, and Beie H -J 1992 *Thin Solid Films.* **207** 288
[7] Kishino T, Shinoda R, Shimizu K, Saioh Y, Ishikawa N, Okamoto Y, Hori F, Matsui T, and Iwase A 2014 *Jpn. J. Appl. Phys.* **53** 05FC07
[8] Shimizu K, Kosugi S, Tahara Y, Yasunaga K, Kaneta Y, Ishikawa N, Hori F, Matsui T, and Iwase A 2012 *Nucl. Instrum. Methods Phys. Res., Sect.* B **286** 291
[9] Nambissan P M G 2013 *J. Phys.: Conf. Ser.* **443**, 012040
[10] Liu Y, Lockman Z, Aziz A, and MacManus-Driscoll J 2008 *J. Phys.: Condens. Matte*r **20** 165201
[11] Maensiri S, Masingboon C, Laokul P, Jareonboon W, Promarak V, Anderson P L, and Seraphin S 2007 *Cryst. Growth Des.* **7** 950
[12] Serpone N 2006 *J. Phys. Chem.* B **110** 24287
[13] Bulfin B, Lowe A J, Keogh K A, Murphy B E, Lübben O, Krasnikov S A, and Shvets I V 2013 *J. Phys. Chem. C* **117** 24129

Journal of Physics: Conference Series **674** (2016) 012016 doi:10.1088/1742-6596/674/1/012016

Defect studies of zirconia implanted by high energy Xe ions

O Melikhova[1], J Čížek[1], I Procházka[1], P Hruška[1], V A Skuratov[2], T E Konstantinova[3] and I A Danilenko[3]

[1]Charles University in Prague, Faculty of Mathematics and Physics, V Holešovičkách 2, 18000 Praha 8, Czech Republic
[2]Flerov Laboratory of Nuclear Reactions, Joint Institute for Nuclear Research, Dubna, Moscow region, 141980 Russia
[3]Galkin Donetsk Institute for Physics and Engineering, National Academy of Sciences of Ukraine, Luxemburg Street 72, 83114 Donetsk, Ukraine

E-mail: oksivmel@yahoo.com

Abstract. In the present work positron lifetime spectroscopy was employed for characterization of radiation-induced defects in yttria stabilized zirconia (YSZ) implanted by 167 MeV Xe ions. Positron lifetime data were interpreted with aid of *ab-initio* theoretical modelling of defects in YSZ lattice. Damage caused by Xe implantation was investigated in two YSZ samples with different microstructure: (i) single crystal and (ii) sintered ceramic. The virgin YSZ single crystal exhibits single component spectrum with lifetime of ≈ 180 ps. Similar lifetime component was found also in the virgin sample of sintered YSZ ceramic. Since this lifetime is significantly higher than the YSZ bulk lifetime the virgin YSZ crystal and the sintered ceramic both contain vacancy-like defects. Xe implantation leads to appearance of additional defect component with longer lifetime ≈ 370 ps which comes obviously from vacancy clusters formed by agglomeration of irradiation induced vacancies. A broad absorption band with peak absorption at ≈ 518 nm was found in Xe-implanted crystal by optical measurements.

1. Introduction

Zirconia (ZrO_2) is used in a wide range of high temperature applications due to its high melting point, low thermal conductivity, high strength and good fracture toughness [1]. Moreover, zirconia is known as radiation hard material resistant against radiation damage [2]. However, the pure zirconia, which is monoclinic at room temperature, undergoes a transformation to the tetragonal phase at 1100°C temperature. Such a transformation is connected with a considerable volume shrinkage leading even to a creation of cracks within ZrO_2 structure [3]. Obviously, this effect inhibits any use of the pure zirconia in high temperature applications. Thus, a stabilization of the high-temperature ZrO_2 phase is required to make a full use of advantageous zirconia features in industry. This can be achieved by an addition of trivalent yttrium oxide Y_2O_3 (yttria) which forms a solid solution in the ZrO_2 lattice. Materials, in which the stabilization of the high-temperature phase is attained by means of an yttria addition, are conventionally called yttria-stabilized zirconia (YSZ) [4]. An amount of ≈ 3 mol. % of the Y_2O_3 is sufficient to stabilize the tetragonal ZrO_2 phase down to ambient temperature. The high temperature cubic phase of YSZ is stabilized at room temperature by adding more than ≈ 8 mol. % of the Y_2O_3 stabilizer to zirconia.

The present work presents characterization of radiation-induced defects in YSZ implanted by heavy Xe ions with energy of 167 MeV. To examine the effect of microstructure on the radiation hardness two YSZ samples with different microstructure were compared: (i) single crystal and (ii) sintered ceramic. Positron annihilation spectroscopy (PAS) was employed for characterization of defects created by Xe implantation. PAS investigations were combined with optical measurements.

2. Experimental

A ZrO_2 + 10 mol.% Y_2O_3 (110) single crystal with cubic structure and tetragonal ZrO_2 + 3 mol.% Y_2O_3 ceramic sintered at 1350°C were studied. The ceramics exhibits the mean grain size of ~ 250 nm. In the following text these samples are denoted YSZ10 and YSZ3, respectively. Both samples were implanted by Xe^{26+} ions with energy of 167 MeV up to a fluence of 3×10^{12} cm^{-2}. The Xe irradiations were made at IC-100 FLNR cyclotron in JINR, Dubna. The temperature of the samples during Xe irradiation did not exceed 50 °C.

PAS investigations were performed using a $^{22}NaCO_3$ positron source with activity of 1 MBq deposited on 2 μm thick mylar foil. A digital spectrometer [5] with time resolution of 145 ps (FWHM of resolution function) was employed for positron lifetime (LT) spectroscopy. At least 10^7 positron annihilation events were collected in each LT spectrum which was decomposed into exponential components by a maximum likelihood code [6]. Contribution of annihilations from positron source was derived on the basis of reference LT measurements with a well-annealed α-Fe sample with intensity adjusted for YSZ according to the procedure suggested in Ref. [7]. The source contribution consisted of two components with lifetimes 368 ps and 1.5 ns and corresponding relative intensities 7 and 1%. Coincidence Doppler broadening (CDB) investigations were carried out using a digital spectrometer [8] equipped with two HPGe detectors. At least 10^8 positron annihilation events were accumulated in each CDB spectra.

3. Theoretical calculations

Ab-initio calculations of positron lifetimes were performed within the so-called standard scheme [9] employing the atomic superposition (ATSUP) method [10]. The electron–positron correlation was treated using the local density approximation (LDA) according to the parametrization by Boroński and Nieminen [11] taking into account incomplete positron screening [12] considering the high-frequency dielectric constant as $\varepsilon_\infty = 4.62$. The calculations were performed for 768 atom based supercells. Point defects (vacancies and vacancy clusters) were modeled by removing the corresponding number of Zr or O atoms from the supercell.

4. Results and discussion

4.1. Virgin samples

Results of LT investigations for the virgin samples are collected in Table 1. The YSZ10 sample exhibits a single component LT spectrum with lifetime of ≈ 179 ps which is significantly higher than the zirconia bulk lifetime [13]. Hence, the YSZ10 exhibits high concentration of defects already in the virgin state and virtually all positrons are annihilated in the trapped state. The lifetime of 179 ps can be attributed to positrons trapped at defects with open volume comparable to zirconium vacancy (V_{Zr}) [13]. High density of defects in the as-grown YSZ single crystal was confirmed also by positron back-diffusion measurement performed using a variable energy slow positron beam which yielded very short positron diffusion length of ≈ 9 nm only [13]. Note that theoretical *ab-inito* calculations performed in Ref. [13] revealed that V_{Zr} are deep positron traps but oxygen vacancies (V_O) are incapable of positron trapping at room temperature.

The virgin YSZ3 sample exhibits more complicated LT spectrum consisting of three components listed in Table 1. The short-lived component with lifetime $\tau_1 \approx 40$ ps comes from free positrons not trapped at defects. The dominating component with lifetime $\tau_2 \approx 176$ ps (relative intensity $I_2 \approx 88\%$)

can be attributed to vacancy like defects similar to those in the virgin YSZ single crystal. The long-lived component with lifetime $\tau_3 \approx 320$ ps represents a contribution of positrons trapped at larger point defects with open volume corresponding to several vacancies. As shown in Ref. [13] these larger point defects are located at triple points, i.e. intersections of three of more grain boundaries. Considerable intensity ($I_3 \approx 7\%$) of positrons trapped at triple points is due very small grain size (≈ 250 nm) of the sintered ceramic. The effect of sintering on microstructure of YSZ ceramic was studied in Ref. [14] using variable energy slow positron beam. The density of defects strongly decreases during sintering which was revealed by lowered *S* parameter and also by more gradual decrease of *S* from the surface to bulk. However, determination of the mean positron diffusion length was found to be difficult since the microstructure of sintered ceramic became non-homogeneous and defect density decreased from the surface towards bulk [14]. The positron diffusion length of ≈15 nm estimated in the sub-surface region of the ceramic sintered at 1350°C can be considered as a lower bound. The mean positron diffusion length in bulk is definitely higher. Hence assuming $L_+ \approx 15$ nm as a lower bound of the positron diffusion length and the mean grain size of $d \approx 250$ nm the fraction of positrons which can reach grain boundaries containing triple points by diffusion can be roughly estimated as $1-(d-L_+)^2/d^2 \approx 12\%$. Taking into account that grain boundaries are defects with large cross-section for positron trapping one can expect that larger portion of these positrons will be trapped at triple points. This is in accordance with considerable intensity $I_3 \approx 7\%$ of positrons trapped at triple points resolved here in the LT spectrum of the virgin YSZ3 sintered ceramic. Note that for nanocrystalline YSZ samples which exhibit even smaller grain sizes the intensity of positrons trapped at triple points was correspondingly higher [13].

Table 1. Results of LT investigations: lifetimes τ_i and relative intensities I_i of the components resolved in LT spectra. The standard deviations are given in parentheses in the units of the last significant digit.

sample	τ_1 (ps)	I_1 (%)	τ_2 (ps)	I_2 (%)	τ_3 (ps)	I_3 (%)
YSZ10: ZrO_2 + 10 mol.% Y_2O_3 (110) single crystal						
virgin	-	-	178.5(1)	100	-	-
Xe-implanted	-	-	178.1(7)	95.1(6)	370(10)	4.9(5)
YSZ3: ZrO_2 + 3 mol.% Y_2O_3 sintered ceramic						
virgin	40(10)	5(2)	176(2)	88(1)	322(9)	7.0(6)
Xe-implanted	-	-	177.6(8)	90.5(7)	355(8)	9.5(6)

4.2. Xe-implanted samples

Implantation of 167 MeV Xe ions into zirconia target was simulated by the SRIM code [15] and results are plotted in Fig. 1. The depth profile of Xe ions is plotted by solid line in the figure while shaded area shows the depth profile of vacancies created by implantation. The stopping depth of Xe ions into zirconia is $z_{Xe} \approx 10.8$ nm. For the ion fluence of 3×10^{12} cm^{-2} used here the damage dose at the peak of the damage distribution is ≈ 0.019 dpa. In Fig. 1 the depth profile of Xe ions is compared with the implantation profile of positrons emitted by ^{22}Na radioisotope (dashed line). The positron implantation profile can be described by exponential function [16]

$$p(z) = \frac{1}{z_0}\exp\left(-\frac{z}{z_0}\right), \tag{1}$$

where z is the depth from the surface. The mean positron penetration depth z_0 can be expressed as

$$z_0 = \frac{E_{\max}^{1.4}}{16\rho}. \quad (2)$$

The symbol ρ in the above equation stands for the density of the sample (expressed in g cm^{-3}) and $E_{\max}$ is the end-point energy of the emitted positrons (expressed in MeV). The mean penetration depth of the positrons emitted by the ^{22}Na radioisotope ($E_{\max}$ = 0.545 MeV) into zirconia is 48 μm. The probability P that a positron emitted by ^{22}Na source is stopped in the layer modified by Xe implantation is obtained by integrating the positron implantation profile

$$P = \int_0^{z_{Xe}} \frac{1}{z_0} \exp\left(-\frac{z}{z_0}\right) dz. \quad (3)$$

For Xe-implanted YSZ specimens Eq. (3) yielded P = 20%. Hence, considerable fraction of positrons is annihilated in the layer affected by Xe implantation.

The YSZ10 single crystal implanted by Xe ions exhibit two-component LT spectrum, see Table 1. The dominating contribution comes again from V_{Zr} which represented the dominating kind of positron traps also in the virgin crystal. A new component with lifetime $\tau_3 \approx$ 370 ps which appeared in LT spectrum of the implanted crystal can be attributed to positrons trapped at larger point defects with open volume corresponding to several vacancies. Bombarding high energy Xe ions create high number of vacancies in zirconia, see Fig. 1. SRIM simulations revealed that *on average* ~ 19000 vacancies (both V_{Zr} and V_O) are created by each Xe ion. Since implantation is performed at room temperature most of these vacancies are annealed, i.e. they either recombine with interstitials or disappear by diffusion to sinks at grain boundaries. However, some irradiation-induced vacancies agglomerate into vacancy clusters. Because of saturated trapping monovacancies created by implantation cannot be distinguished from those present already in the virgin crystal. But irradiation-induced vacancy clusters can be distinguished and led to rise of the long-lived component with the lifetime τ_3 in the implanted crystal. It has to be noted that annealing of vacancies is not considered in SRIM simulations, i.e. Fig. 1 shows the concentration of vacancies *created* during implantation but not the concentration of vacancies *survived* in the irradiated sample at room temperature.

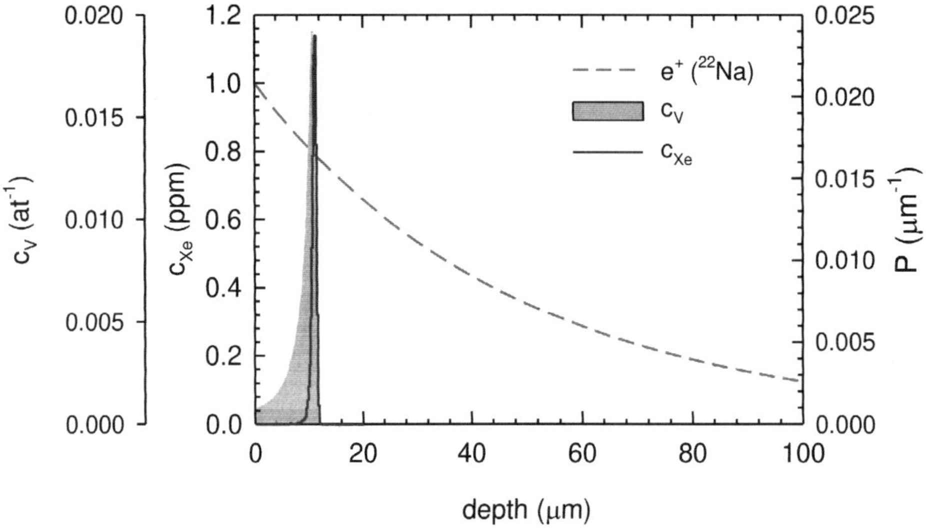

Figure 1. Results of SRIM simulations for 167 MeV Xe ions implantation into zirconia: the depth profiles of implanted Xe ions (c_{Xe}, blue solid line) and vacancies created by implantation (c_V, filed grey area). Dashed line shows the implantation profile of positrons emitted by ^{22}Na radioisotope.

Ab-initio theoretical calculations were employed for determination of the average size of vacancy clusters created by Xe irradiation. Calculated positron lifetimes for vacancy clusters of various sizes are plotted Fig. 2 as a function of the effective cluster 'size' calculated as $n_{V_{Zr}} d_{Zr} + m_{V_O} d_O$, where $n_{V_{Zr}}$ and m_{V_O} is the number of V_{Zr} and V_O in the cluster and d_{Zr} and d_O is the diameter of Zr and O atom, respectively. From inspection of Fig. 2 it becomes clear that the lifetime $\tau_3 \approx 370$ ps observed in the Xe-implanted crystal corresponds to vacancy clusters consisting of 6 V_{Zr} and 6 V_O ($6V_{Zr}$+$6V_O$). Note that the lifetime for vacancy clusters in Xe-implanted crystal is remarkable higher than the lifetime of 312 ps reported for YSZ implanted by 2 MeV He ions [17]. The lifetime of 312 ps measured in the He implanted YSZ corresponds to vacancy clusters consisting of 4 V_{Zr} and 4 V_O ($4V_{Zr}$+$4V_O$), see Fig. 2. Hence, irradiation by heavy Xe ions with high energy created larger vacancy clusters than irradiation by lighter and lower energy He ions.

The YSZ3 ceramic implanted by Xe ions exhibits saturated positron trapping at V_{Zr} (lifetime $\tau_2 \approx$ 178 ps) and larger point defects characterized by the lifetime $\tau_3 \approx 355$ ps, see Table 1. Disappearance of the free positron component testifies that the concentration of defects in the Xe-implanted ceramic increased. Moreover, the lifetime τ_3 is enhanced and the intensity I_3 is increased with respect to the virgin ceramic. This testifies that in the Xe-implanted ceramic the long-lived component with lifetime τ_3 represents a contribution of positrons trapped at triple points (lifetime ≈ 320 ps) which are present already in the virgin sample and also vacancy clusters (lifetime ≈ 370 ps) created by Xe irradiation.

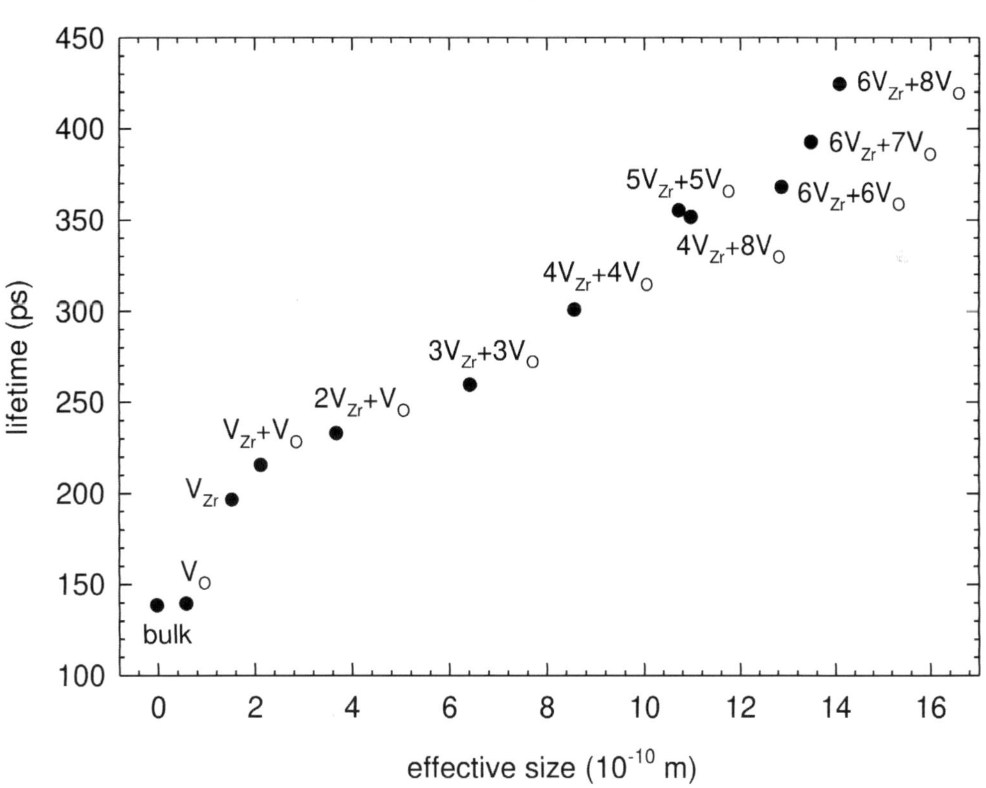

Figure 2. Results of *ab-inito* theoretical calculations of positron lifetimes for various defects in zirconia. Labels show the number of Zr-vacancies (V_{Zr}) and O-vacancies (V_O) forming the cluster. Positron lifetimes are plotted versus the effective 'size' of defects, see text.

Results of CDB investigations are presented in Fig. 3 as ratio curves with respect to well annealed pure Zr. The ratio curves for the virgin and Xe-implanted YSZ10 crystal and YSZ3 ceramic are compared in Fig. 3a and 3b, respectively. The prominent feature of all ratio curves measured is a pronounced peak at $p \approx 15 \times 10^{-3}$ m_0c which represents a contribution of positrons annihilated by 2*p* oxygen electrons [13]. The ratio curves for YSZ10 and YSZ3 samples are very similar and the contribution of positrons annihilated by oxygen electrons is significantly higher than in YSZ nano-powders measured in Ref. [13]. This is in agreement with LT investigations which identified V_{Zr} as dominant positron traps. Since oxygen anions are the nearest neighbours of V_{Zr} the high momentum part of the momentum distribution of positrons trapped at V_{Zr} contains prominent contribution from positrons annihilated by oxygen electrons. One can see in Fig. 3 that the ratio curves for YSZ10 and YSZ3 samples implanted by Xe ions are almost the same as those for the corresponding virgin materials. This indicates in accordance with LT studies that V_{Zr} remain the dominating positron traps also in the Xe-implanted samples.

Fig. 4a shows optical transmittance of the virgin and Xe-implanted YSZ10 crystal. The virgin crystal is colourless and transparent in the whole spectrum of the visible light. Implantation by Xe ions turned the crystal grey and a broad minimum appeared in the optical transmittance curve. This gives clear evidence that defects providing electron states in the band gap and allowing for absorption of visible light were created by Xe-impantation. The difference between optical transmittance of the Xe-implanted and the virgin YSZ10 crystal is plotted in Fig. 4b. Obviously the Xe-implanted crystal absorbs light in a broad range of wavelengths 350-650 nm. Similar broad absorption band has been reported in a cubic ZrO_2 + 9.5 mol.% Y_2O_3 single crystal implanted by 200 keV Xe^+ ions [18]. However, the absorption band was observed only after irradiation up to the Xe ion fluence of 10^{16} cm^{-2} or higher [18]. Here the absorption band was observed already at significantly lower fluence of 3×10^{12} cm^{-2} due to much higher energy of implanted Xe ions. The broad absorption band was previously observed in X-ray irradiated YSZ sample and has been attributed to two types of defects: (i) V_O with trapped electron (F-type centre) and (ii) hole-trapped oxygen ion (V-type centre) adjacent to yttrium cation [19]. Since Xe implantation produces large number of V_O, explanation of the broad absorption band by irradiation-induced F-type centres is very plausible. Note that the absorption band in Fig. 4 exhibits maximum absorption at 518 nm. This is very similar to the peak of the absorption band of 522 nm observed in Xe-implanted YSZ in Ref. [18] but it is shifted to longer wavelength compared the peak of absorption band in X-ray irradiated YSZ (465 nm) [19] and neutron irradiated YSZ (470 nm) [20]. It has been suggested that this red shift could be due to aggregation of vacancies to vacancy clusters [18,21]. Our positron lifetime investigations support this hypothesis. Since irradiation by Xe ions leads higher density of V_O per track length compared to X-rays or neutrons used in Refs. [19,20] the agglomeration of V_O into clusters becomes easier.

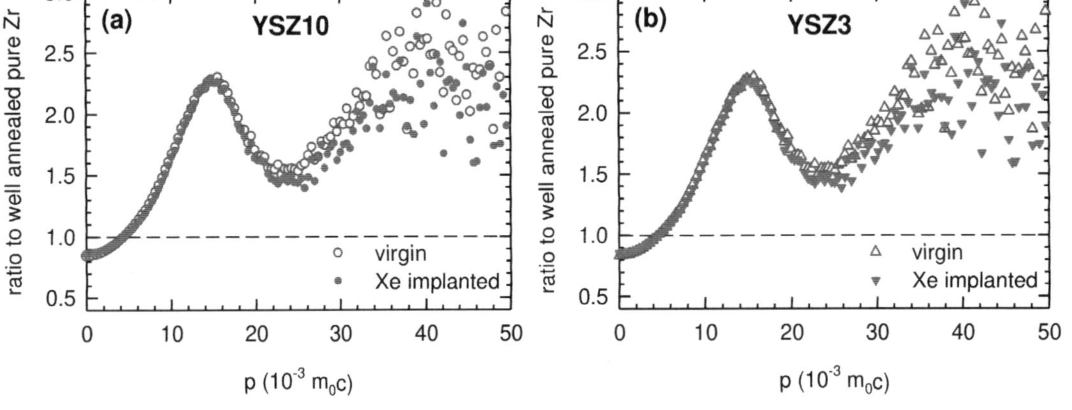

Figure 3. CDB ratio curves (with respect to well annealed pure Zr) for (a) virgin and Xe-implanted YSZ10 crystal and (b) virgin and Xe-implanted YSZ3 ceramic.

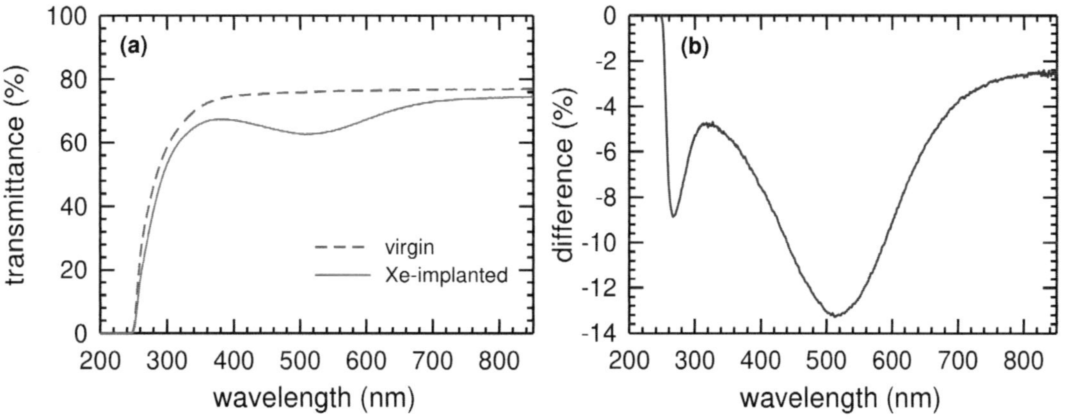

Figure 4. Results of optical measurements of YSZ10 crystal: (a) transmittance of the virgin and Xe-implanted crystal; (b) difference between the optical transmittance curves.

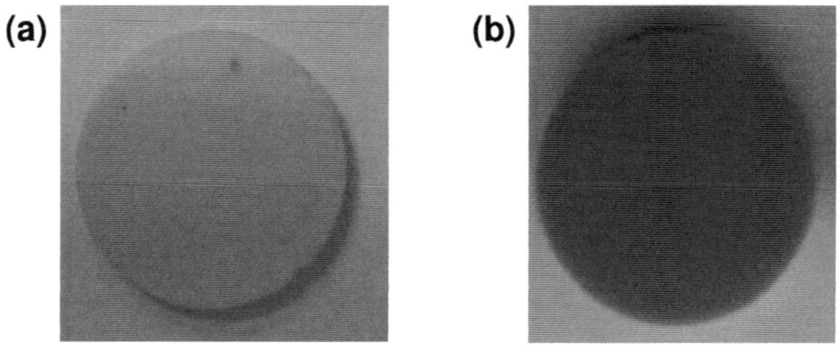

Figure 5. Change of coloration of YSZ3 ceramic after implantation by Xe ions: (a) virgin sample; (b) Xe-implanted sample.

Change of coloration caused by Xe implantation was observed also in YSZ3 ceramic. The virgin ceramic was white as shown Fig. 5a but after implantation by Xe ions it turned pink, see Fig. 5b. Similarly to YSZ10 crystal the change of colour of Xe-implanted YSZ3 ceramic is caused by irradiation-induced colour centres.

5. Conclusions

A cubic ZrO_2 + 10 mol.% Y_2O_3 single crystal and tetragonal ZrO_2 + 3 mol.% Y_2O_3 ceramic were implanted by 167 MeV Xe ions. Defects in the virgin and Xe-implanted samples were characterized by positron annihilation spectroscopy. Both the single crystal and the ceramic sample exhibit V_{Zr} as dominating positron traps in the virgin state. The virgin samples contain most probably also V_O but they are not able to confine positrons. Irradiation by Xe ions introduced larger point defects characterized by positron lifetime of $\approx$ 370 ps and identified as clusters consisting of 6 V_{Zr} and 6 V_O. Xe-implantation introduced colour centres and changed colour of both the single crystal (colourless to grey) and the ceramic (white to pink). Measurement of optical transmittance of Xe-implanted single crystal revealed a wide absorption band (350-650 nm) which is likely caused by irradiation-induced F-type centres.

Acknowledgements

This work was supported by the Czech Science Agency (project P108-11-1396).

Journal of Physics: Conference Series **674** (2016) 012016 doi:10.1088/1742-6596/674/1/012016

References

[1] Li P, Chen I W, James E and Hahn J E P 1993 *Phys. Rev.* B **48** 10063

[2] van Duppen P, Decrock P, Huyse M, Delbar Th, Galster W, Leleux P, Lienard I, Lipnik P, Loiselet M, Michotte C, Ryckewaert G, Vervier J, Duhamel P and Vanhorenbeeck J 1992 *Nucl. Instr. Meth.* B **70** 393

[3] Evans A G and Cannon R M 1986 *Acta Metall.* **34** 761

[4] Yanagida H, Koumoto K, and Miyayama M 1996 *The Chemistry of Ceramics* (Chicester, John Wiley & Sons)

[5] Bečvář F, Čížek J, Procházka I and Janotová J 2005 *Nucl. Instrum. Methods* A **539** 372

[6] Procházka I, Novotný I, Bečvář F 1997 *Mater. Sci. Forum* **255-257** 772

[7] Surbeck H 1977 *Helv. Phys. Acta* **50** 705

[8] Čížek J, Vlček M and Procházka I 2010 *Nucl. Instr. Meth. in Phys. Research* A **623** 982

[9] Puska M J and Nieminen R M 1994 *Rev. Mod. Phys.* **66** 841

[10] Puska M J and Nieminen R M 1983 *J. Phys.* F **13** 333

[11] Boroński E and Nieminen R M 1986 *Phys. Rev.* B **34** 3820

[12] Puska M J, Mäkinen S, Manninen M and Nieminen R M 1989 *Phys. Rev.* B **39** 7666

[13] Čížek J, Melikhova O, Procházka I, Kuriplach J, Kužel R, Brauer G, Anwand W, Konstantinova T E and Danilenko I A 2010 *Phys. Rev.* B **81** 024116

[14] Procházka I, Čížek J, Melikhova O, Anwand W, Brauer G, Konstantinova T E and Danilenko I A 2013 *Mater. Sci. Forum* **733** 236

[15] Ziegler J F, Ziegler M D and Biersack J P 2010 *Nucl. Instr. Meth. in Phys. Research B* **268** 1818

[16] Hautojärvi P and Corbel C 1995 *Positron Spectroscopy of Solids Proc. Int. School of Physics 'Enrico Fermi'* ed A Dupasquier and A P Mills (Amsterdam: IOS Press) p 491

[17] Grynszpan R I, Brauer G, Anwand W, Malaquin L, Saudé S, Vickridge I and Briand E 2007 *Nucl. Instr. Meth. in Physics Research* B **261** 888

[18] Zhu S, Zu X T, Xiang X, Wang Z G, Wang L M and Ewing R C 2003 *Nucl. Instr. Meth. in Physics Research* B **206** 1092

[19] Orera V M, Merino R I, Chen Y, Cases R and Alonso P J 1990 *Phys. Rev.* B **42** 9782

[20] Savoini B, Cáceres D C, Vergara I, González R and Muñoz Santiuste J E 2000 *J. Nucl. Mater.* **277** 199

[21] Grinyaev S N, Konusov F V and Lopatin V V 2002 *Phys. Solid State* **44** 275

Hole size distributions in cardo-based polymer membranes deduced from the lifetimes of ortho-positronium

Y Kobayashi[1,3], A Kinomura[1,6], S Kazama[2], K Inoue[3], T Toyama[3], Y Nagai[3], K Haraya[4], H F M Mohamed[5], B E O'Rourke[1], N Oshima[1], R Suzuki[1,7]

[1]Research Institute for Measurement and Analytical Instrumentation, National Institute of Advanced Industrial Science and Technology, Tsukuba, Ibaraki 305-8568, Japan
[2]Nippon Steel & Sumitomo Metal Corporation, Technical Development & Planning Division, 6-1 Marunouchi, 2 chome, Chiyoda-ku, Tokyo 100-8701, Japan
[3]The Oarai Center, Tohoku University, Oarai, Ibaraki 311-1313, Japan
[4]Research Institute for Chemical Process Technology, National Institute of Advanced Industrial Science and Technology, Tsukuba, Ibaraki 305-8565, Japan
[5]Physics Department, Faculty of Science, Minia University, P. O. Box 61519, Minia, Egypt

E-mail:r-suzuki@aist.go.jp

Abstract. To clarify the free volume size distributions of the cardo-based polymer membranes, where ortho-positronium (o-Ps) undergoes pick-off annihilation, the o-Ps lifetime distributions were analyzed by the LT9 programme. It was found that the cardo-based polysulfone membrane has much narrower o-Ps lifetime/hole size distributions than the cardo-based polyimide membranes with the 2,2-bis(3,4-dicarboxyphenyl)hexafluoropropane dianhydride (6FDA) moiety. Further, the lifetime/hole size distributions of the cardo-based polymer membranes are appreciably broadened with increasing temperature. This suggests that in these membranes there are holes not only of different sizes but also of different thermal expansion coefficients. It is also shown that in a membrane with a wider hole size distribution the average o-Ps lifetime tends to be longer than would be expected from the correlation between the o-Ps lifetime and the total free volume for common polymers.

1. Introduction

Introduction of a bulky cardo moiety into a polymer main chain is expected to hinder the molecular packing. The free volume model of gas diffusion states that the diffusivity of gas molecules in polymers depends on intermolecular open space often called free volume [1,2]. According to this model, sparse molecular packing with large free volume is advantageous to higher gas diffusivity. Therefore cardo-based polymers arouse interest for their application in gas separation membranes [3].

Positron annihilation lifetime spectroscopy (PALS) has been extensively used to study the nanostructure of amorphous polymers [4,5]. Some of the positrons injected into a polymer in many cases form the bound state with an electron, positronium (Ps). The experimental positron lifetime

[6]Present address: Research Reactor Institute, Kyoto University, Kumatori-cho, Sennan-gun, Osaka 590-0494, Japan
[7]To whom any correspondence should be addressed.

spectrum is usually resolved into three exponentially decaying components with lifetimes τ_1, τ_2, τ_3 ($\tau_1 < \tau_2 < \tau_3$) and the corresponding relative intensities I_1, I_2, I_3. The three lifetimes are attributed to the self-annihilation of spin parallel para-positronium (p-Ps), annihilation of non-Ps positrons with surrounding electrons, and pick-off annihilation of o-Ps, in which the positron in o-Ps undergoes annihilation with one of the surrounding spin antiparallel electrons. According to Tao [6], Eldrup and others [7], the pick-off annihilation lifetime of o-Ps can be related to the size of a hole for o-Ps localization. Under a simplified assumption that the holes in a polymer are of spherical shape, the relationship between the pick-off annihilation lifetime (τ_3) of o-Ps and the average radius (R) of the holes is given by

$$\tau_3 = 0.5\left\{1 - \frac{R}{R_o} + \frac{1}{2\pi}\sin\left(\frac{2\pi R}{R_o}\right)\right\}^{-1} \quad (1)$$

where R_0 ($= R + \Delta R$) is the radius of an infinitely deep square well potential and ΔR (= 0.166 nm) is the measure of the overlap of the Ps wave function with electrons on the walls of the open spaces. The hole volume V_h in nm^3 is obtained as

$$V_h = \frac{4\pi}{3}R^3$$

This model, in spite of its simplicity, has been successfully applied to the study of a variety of functional polymers [8].

Recently we applied PALS to a series of cardo-based polyimide and polysulfone membranes [9]. It was found that acid anhydride moiety of polyimide membranes has a strong impact on Ps formation in the sense that no Ps forms in most of the polyimides with pyromellitic dianhydride and 3,3',4,4'-biphenyltetracarboxylic dianhydride moieties with high electron affinity, whereas favourable amounts of Ps are formed in those with 2,2-bis(3,4-dicarboxyphenyl) hexafluoropropane dianhydride (6FDA) moiety with lower electron affinity. Favourable Ps formation was also observed for the cardo-based polysulfone membrane. It has been reported that Ps formation in polysulfones is little influenced by the presence of chemical groups or additives with high electron affinity due to the so-called anti-inhibition effect of $-SO_2-$ groups on Ps formation [10,11]. Furthermore we studied the connectivity between the average hole size of cardo-based polymer membranes deduced by the Tao-Eldrup model and the gas diffusion coefficients. Logarithmic plots of the diffusion coefficients of O_2 and N_2 versus reciprocal average hole size for cardo-based polymers, based on the free volume theory of diffusion, are appreciably shifted from the previously reported correlations for common polymers. The shifts likely originate from the strong tendency of o-Ps staying in larger holes, which is possibly associated with wide hole size distributions of the cardo-based polymer membranes. This paper focuses on the o-Ps lifetime/hole size distributions of the three cardo-based polymer membranes with favourable Ps formation.

2. Experimental

Figure 1 shows the chemical structures of cardo-based polymers studied [9]. The polymers have bis (phenol)fluorene, bis(aniline)fluorene or bis(xylidine)fluorene as the cardo moiety. Cardo polysulfone was supplied by Nippon Steel Corporation. Cardo-polyimides were synthesized by solution polycondensation reactions [3]. The abbreviations of the polymers follow the notation of X-Y-Z, where X means the type of the polymer; X = PI for polyimide and X = PSF for polysulfone. Y and Z are the abbreviations of bifunctional monomers: Y = BXFL for bis(xylidine)fluorene, Y = BAFL for bis(aniline)fluorene, Y = BPFL for bis (phenol)fluorene, Z= 6FDA for 2,2-bis(3,4-dicarboxyphenyl) hexafluoropropane dianhydride, Z = DCPS for 4,4'-dichlorodiphenyl sulfone. The total free volume V_f of polymer was calculated from the following

$$V_f = V - V_0$$

where V is the specific volume and V_0 the occupied volume. The occupied volume was estimated from the van der Waals volume (V_w) according to the relationship $V_0 = 1.3V_W$ [12].

Positron lifetime measurements were performed in a temperature range from 323 K down to 10 K with a fast-fast analogue spectrometer as previously described [9]. The lifetime spectra were analyzed into three components with the LT9 programme [13], assuming log normal distribution of the o-Ps annihilation rates (reciprocal lifetimes). The analysis provided the average o-Ps lifetime (τ_{av}), dispersion of o-Ps lifetime distributions and the relative intensity as well as two shorter discrete lifetimes and their relative intensities. From the obtained dispersions, o-Ps lifetime distributions at different temperatures were computed. At least 3 million counts were collected for each lifetime spectrum, which enabled the reliable analysis of the o-Ps lifetime distribution. Further, by applying the Tao-Eldrup model, hole volume distributions were obtained.

PSF-BPFL-DCPS

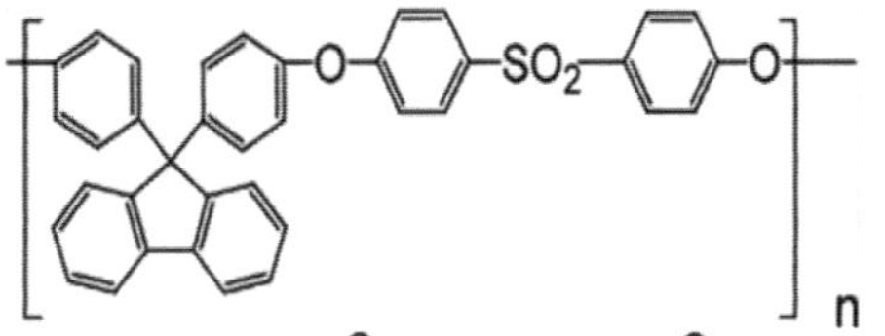

PI-BAFL-6FDA

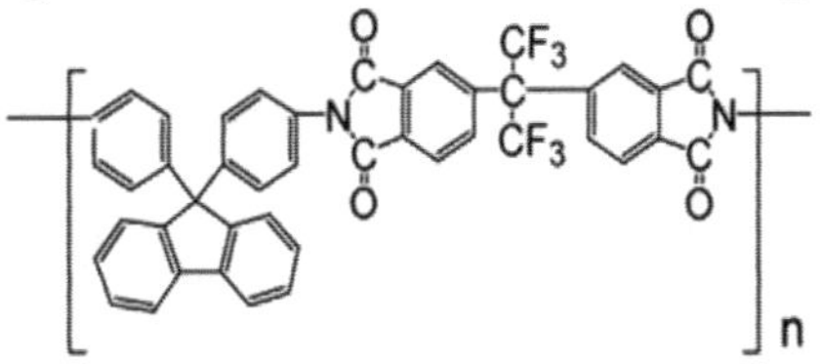

PI-BXFL-6FDA

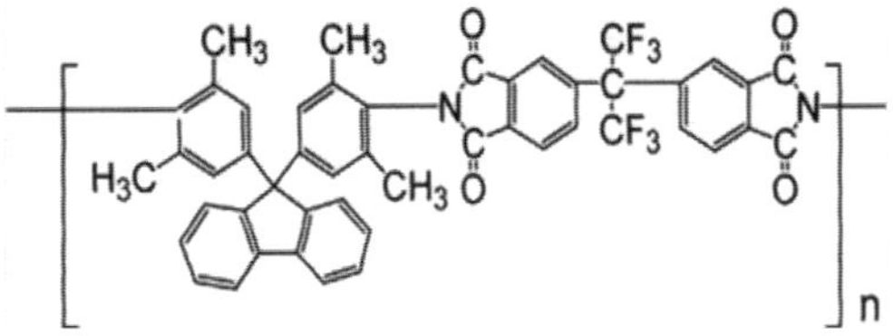

Figure 1. Chemical structures of PSF-BPFL-DCPS, PI-BAFL-6FDA and PI-BXFL-6FDA [9].

3. Results and discussion

In LT9 log normal distribution is assumed for o-Ps annihilation rates λ ($=1/\tau_3$).

$$\alpha_3(\lambda) = -(2\pi)^{-\frac{1}{2}}\sigma_3^{-1}\exp[-\frac{\{(\ln(\lambda) - \ln(\lambda_{30})\}^2}{2\sigma_3^2}]\lambda^{-1}d\lambda$$

where σ_3 is the dispersion of the distribution and λ_{30} is the peak annihilation rate of o-Ps. From this, the o-Ps lifetime distribution is given by

$$\alpha_3(\tau_3) = -(2\pi)^{-\frac{1}{2}}\sigma_3^{-1}\exp[-\frac{\{(\ln(\lambda)-\ln(\lambda_{30})\}^2}{2\sigma_3^{\,2}}]\lambda^{-1}d\lambda\frac{d\tau_3}{d\lambda}$$

$$= (2\pi)^{-\frac{1}{2}}\sigma_3^{-1}\exp[-\frac{\{(\ln(\lambda)-\ln(\lambda_{30})\}^2}{2\sigma_3^{\,2}}]\lambda d\tau_3$$

Further the hole volume distribution is expressed as

$$\alpha_3(V) = (2\pi)^{-\frac{1}{2}}\sigma_3^{-1}\exp[-\frac{\{(\ln(\lambda)-\ln(\lambda_{30})\}^2}{2\sigma_3^{\,2}}]\lambda^{-1}d\lambda\frac{dV}{d\lambda}$$

where from equation (1),

$$\frac{dV}{d\lambda} = -\frac{4\pi R^2(R+R_0)^2}{2R_0}\left(1-\cos\frac{2\pi R}{R+R_0}\right)^{-1}$$

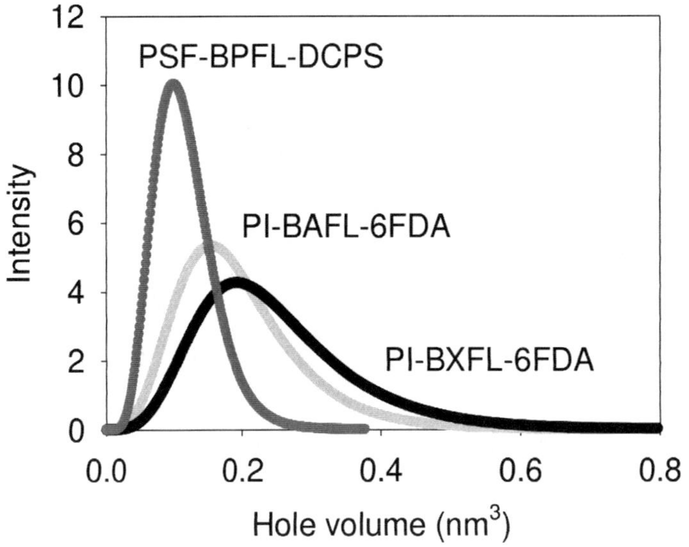

Figure 2. Hole volume distributions of the three cardo membranes at 298K deduced from o-Ps lifetime distributions.

Figure 2 shows the distributions of hole volumes for PI-BXFL-6FDA, PI-BAFL-6FDA and PSF-BPFL-DCPS at 298 K obtained by the above equations. One can see that the distribution is broader in the order of PI-BXFL-6FDA > PI-BAFL-6FDA >> PSF-BPFL-DCPS. In figure 3 is shown the temperature dependence of the peak positions and full widths at half maximum (FWHM's) of the hole volume distributions deduced from the o-Ps lifetimes for the three membranes. For all the membranes peak positions gradually shift to higher values with increasing temperature, which is reasonably attributed to the thermal expansion of the open spaces for Ps localization. Larger hole volumes in PI-BXFL-6FDA than PI-BAFL-6FDA at all the temperatures studied may be attributed to the presence of the more bulky methyl substituted BXFL moiety in the former membrane [9]. In figure 3 we also note that at every temperature the distribution is broader (FWHM is larger) in the order of PI-BXFL-6FDA > PI-BAFL-6FDA >> PSF-BPFL-DCPS. It is conjectured that in the former two membranes because of the stiff chains of polyimide the structures around large intermolecular spaces are not able to relax sufficiently and molecular packing is hindered, resulting in the wide hole size distributions peaked at larger hole sizes. This is not surprising because Shimazu et al [14] reported that o-Ps lifetimes in other polyimides with 6FDA moieties are also widely distributed at room temperature. The lifetime/hole size distributions of the membranes, in particular, PI-BXFL-6FDA and PI-BAFL-6FDA are considerably broadened with increasing temperature, which suggests that in these cardo-based membranes there are holes of not only different sizes but also of different thermal expansion coefficients. In PSF-BPFL-DCPS, molecular chains are more flexible than polyimides [3, 9] and hence the distributions are much narrower and the holes are smaller.

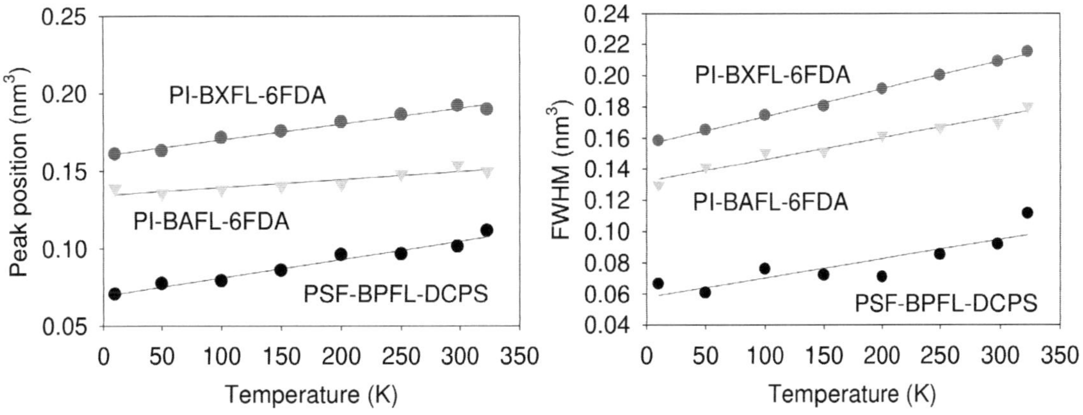

Figure 3. Peak positions (left) and FWHM's (right) of the hole volume distributions of the three cardo-based membranes as a function of temperature.

Figure 4 shows the plot of the average o-Ps lifetime versus total free volume for common polymers such as polystyrene, polycarbonate and so forth [15] as well as the three cardo-based polymer membranes at 298 K. For common polymers, where the hole size distributions are not significantly broadened, there is a good correlation between the two quantities; the o-Ps lifetime systematically increases from 0.5 ns to higher values with increasing total free volume. However, the data points for PI-BXFL-6FDA, PI-BAFL-6FDA and PSF-BPFL-DCPS are all deviated upward from the correlation, indicating that the average o-Ps lifetimes in these polymers are appreciably longer than expected from the correlation for the common polymers. The deviation is the largest for PI-BXFL-6FDA and the smallest for PSF-BPFL-DCPS. Therefore, the broader the lifetime/hole size distribution in figure 2, the larger the deviation from the correlation for the common polymers in figure 4.

According to molecular dynamics simulation of polymers [14, 16], the holes in the amorphous region of polymers are not closed, with opening and well connected to each other. Therefore, in a polymer with a wide hole size distribution, Ps formed in a smaller hole may migrate quantum mechanically to a neighbouring larger hole through connecting channels. In a polymer with a particularly wide hole size distribution such as PI-BXFL-6FDA, such a process is repeated a number of times until o-Ps gets trapped in the largest available hole. Once trapped in the largest available hole, o-Ps cannot go back to the smaller hole because of reduced zero-point energy in it, which results in the average o-Ps lifetime considerably longer than that expected from the average size of the actual hole size distribution. This is likely the reason why the data points for PI-BXFL-6FDA, PI-BAFL-6FDA and PSF-BPFL-DCPS are deviated from the correlation for common polymers in figure 4.

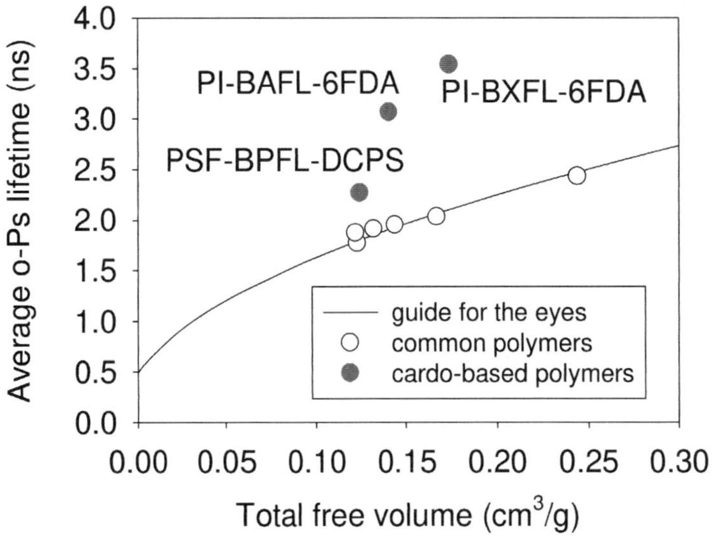

Figure 4. Correlation between the average o-Ps lifetimes and the total free volume for common polymers such as polycarbonate, polystyrene, polysulfone, polyethersulfone etc at 298 K. The line shows the guide for the eyes under the assumption that the o-Ps lifetime in a polymer with no free volume is 0.5 ns. Included in the figure are the data for PI-BXFL-6FDA, PI-BAFL-6FDA and PSF-BPFL-DCPS at 298 K.

4. Conclusion

Positron lifetime spectra for three cardo-based polymer membranes, PI-BXFL-6FDA, PI-BAFL-6FDA and PSF-BPFL-DCPS with favourable Ps formation, recorded at different temperatures, were analyzed by the LT9 programme and the o-Ps lifetime/hole size distributions were obtained. At each temperature, the hole size distributions are broader in the order of PI-BXFL-6FDA > PI-BAFL-6FDA >> PSF-BPFL-DCPS. The hole volume distributions of the membranes, in particular, PI-BXFL-6FDA and PI-BAFL-6FDA are considerably broadened with increasing temperature, which suggests that in these cardo-based membranes there are holes not only of different sizes but also of different thermal expansion coefficients. The average o-Ps lifetimes of the three cardo-based membranes at 298 K are considerably longer than expected from the correlation between the o-Ps lifetime and total free volume for common polymers. Quantum mechanical migration of o-Ps from smaller holes to larger holes in a polymer with a wide hole size distribution may be responsible for the appreciable upward deviations of the o-Ps lifetimes from the correlation for common polymers.

Acknowledgement

We are grateful to Prof. J. Kansy of Silesian University (Poland) for his support in the positron lifetime distribution analysis by the LT9 programme.

References

[1] Cohen M H and Turnbull D 1959 *J. Chem. Phys.* **31** 1164
[2] Fujita H 1961 *Fortschr. Hochpolym. Forsch.* **3** 1
[3] Kazama S, Teramoto T and Haraya K 2002 *J. Membr. Sci.* **207** 91
[4] Schrader D M and Jean Y C 1988 *Positron and Positronium Chemistry* (Amsterdam: Elsevier)

[5] Mogensen O E 1995 *Positron Annihilation in Chemistry* (Berlin: Springer-Verlag)
[6] Tao S J 1972 *J. Chem. Phys.* **56** 5499
[7] Eldrup M, Lightbody D and Sherwood J N 1981 *Chem. Phys.* **63** 51
[8] Kobayashi Y 2012 *Defect and Diffusion Forum.* **331** 253
[9] Kobayashi Y, Kazama S, Inoue K, Toyama T, Nagai Y, Haraya K, Mohamed H F M, O'Rourke B E, Oshima N, Kinomura A and Suzuki R 2014 *J. Phys. Chem.* B **118** 6007
[10] Hirata K, Kobayashi Y and Ujihira, Y 1997 *J. Chem. Soc. Faraday Trans.* **93** 139
[11] Kobayashi Y, Mohamed H F M and Ohira A 2009 *J. Phys. Chem.* B **113** 5698
[12] Bondi A 1964 *J. Phys. Chem.* **68** 441
[13] Kansy J 1996 *Nucl. Instrum. Methods in Physics Research* A **374** 235
[14] Shimazu A, Miyazaki T, Katayama S and Ito Y 2003 *J. Polym. Sci. Part* B *Polymer Physics* **41** 308
[15] Kobayashi Y, Haraya K, Kamiya Y and Hattori S, 1992 *Bull. Chem. Soc. Japan* **65** 160
[16] Shimazu A, Goto H, Shintani T, Hirose R, Suzuki R and Kobayashi Y, 2013 *J. Phys. Conf. Ser.* **443** 012050

Premelting as studied by positron annihilation and emission Mössbauer spectroscopies

S V Stepanov[1,2], V M Byakov[2,3], D S Zvezhinskiy[2], G Duplâtre[4], L Yu Dubov[1], P S Stepanov[1], Yu D Perfiliev[5], L A Kulikov[5]

[1] National Research Nuclear University "MEPhI" (Moscow Engineering and Physical Institute), 115409, Moscow, Russia
[2] Institute of Theoretical and Experimental Physics, 117218, Moscow, Russia
[3] D. Mendeleyev University of Chemical Technology of Russia, 125047, Moscow, Russia
[4] Institut Pluridisciplinaire Hubert Curien, CNRS/IN2P3, BP 28 67037 Strasbourg, France
[5] Lomonosov Moscow State University, Chemical Department, Leninskie Gory, Moscow, Russia

stepanov@itep.ru

Abstract. We have estimated a local heating which takes place owing to the ionization energy losses at the terminal part of a fast positron track and at nano-vicinities of the ^{57}Fe Mössbauer nuclei in case of the emission Mössbauer spectroscopy. It is shown that in experiments close to the melting point one may expect local melting near the probe species.

1. Introduction

It is well known that positrons (e^+) as well as positronium atoms (Ps) are convenient probes of the local nanoscale structure in a condensed phase and the radiolytic processes occurring therein [1]. When a fast positron is injected into a molecular medium it loses most of its energy through ionizations. Within a few picoseconds this energy is locally converted into heat. Slightly below the melting point the heat is mainly transformed into a latent heat of fusion. As a result, a local melting of the terminal part of the e+ track becomes possible. The radius of the molten region turns out to be several tens of Å. Consequently, the lifetime of the ortho-Ps atom produced therein, τ_3, reaches its liquid-phase value at a temperature 2-3 degrees below the melting point. Apparently, this is due to the fact that the Ps atom is captured by the molten region of the e+ track, where it forms a nanobubble (typical for liquids), while the bulk of the medium remains frozen [2]. Probably this Ps bubble may survive (as a local cavity), after cooling and solidification of this area.

The sensitivity of e+ and Ps to the presence of structural defects opens a possibility to study the nature of phase transitions and, particularly, the accumulation of these defects at a stage of premelting, by detecting the annihilation radiation. We have carried out such a study in frozen alcohols and water [3]. Our preliminary estimations have shown that defects are accumulated in accordance with the Arrhenius law. The total energy spent for their formation approximately agrees with the latent heat of fusion.

In water, formation of the Ps bubble state in the premelted region is somewhat special. Because of the "negative" pressure within the molten "droplet" (due to the difference of densities between ice and liquid water), the Ps bubble becomes even larger than in bulky liquid water [4].

In this paper we discuss the local heating effect in the terminal part of the fast positron tracks as well as a similar effect which may take place in experiments on the emission Mössbauer spectroscopy (EMS) in frozen aqueous solutions. In the latter case local heating (and probably premelting) occur due to the emission and slowing down of the Auger electrons released by excited ^{57}Fe Mössbauer nuclei [5].

Further, fast radiation-chemical reactions in the Auger-blob (ion-electron recombination, electron localization and scavenging) determine the experimentally observable ratio of the yields of final chemically stable Fe^{3+} and Fe^{2+}. This ratio is important for adequate interpretation of the emission Mössbauer spectra. Reduction of the Fe^{3+} ion to the Fe^{2+} state occurs because of interaction with one of the blob quasifree electrons [6]. In [2] we have discussed the structure of a positron track. It was shown that, in its terminal part, positron ionization losses are most efficient. So there is the highest density of ion-electron pairs and chemically active radiolytic products (radicals), which are able to react with the positronium atom, formed therein. In accordance with radiation-chemical literature this part of the positron track is called the terminal blob. It consists of several tens of ion-electron pairs. Their spatial distribution can be approximated by the Gaussian function

$$c(r,t=0)=N\frac{\exp(-r^2/a^2)}{\pi^{3/2}a^3}, \qquad (1)$$

where $a\approx 40$ Å and $N\approx 30$ [2].

The positron blob formation energy, $W_{blob}\approx 500$ eV, is released within the volume of the e+ blob during a few femtoseconds. Then this energy is transformed into heat (over picoseconds), which leads to an increase of the local temperature of the blob.

A similar scenario occurs in the emission Mössbauer spectroscopy (EMS). In EMS experiments excited Mössbauer $^{57}Fe^*$ nuclei appear as a result of an electron capture by ^{57}Co nuclei (^{57}Co half-life is 272 days). Thus, the ^{57}Co nucleus transforms into a highly excited state, $^{57}Fe^*$ (136 keV), which during approximately 9 ns populates the low-lying Mössbauer nuclear level at 14.4 keV (its half-life is 98 ns).

Just after E-capture, the electronic structure of the Mössbauer iron ion corresponds to that of the initial $^{57}Co^{2+}$ ion, but with an electronic vacancy ("hole") usually on the K-shell. The excitation energy of this electronically excited iron ion may be estimated, knowing the wavelength-Kα lines of the characteristic radiation of iron or cobalt, which are respectively equal to 0.194 nm and 0.179 nm (or 6.4 keV and 6.9 keV in energy units). This hole on the K-shell is filled with one of the electrons from the outer shells. This transition initiates the emission of the cascade of the Auger electrons with a total energy $W_{blob}\approx 6$ keV. The emission of the Auger cascade leads to the formation of a multiply ionized iron ion, $^{57}Fe^{n+}$, where n may reach 8, averaging at 4-6.

However, this emission leads not only to formation of the multi-charged $^{57}Fe^{n+}$ ion, but also to formation of a cloud of many (200-300) ion-electron pairs (H_2O^+, e^-) around this ion. Such a cloud is termed the Auger blob. Its radius is about ~100 Å and its formation time is about 10^{-13} s [5]. Ionization slowing down of the Auger electrons initiates radiolytic processes in the nearest (nano)vicinity of the Mössbauer nucleus (ion). In [6, 7] EMC was applied for investigating these radiolytic processes in frozen aqueous solutions of different acids.

2. Simulations of the spatial distributions of the secondary electrons in the positron blob and the Auger blob

Modeling of the spatial distribution of the secondary electrons in tracks of the ionizing particles (in the positron blob and in the Auger blob) may be carried out with the help of the GEANT4 toolkit for the Monte Carlo simulation of the passage of particles through matter [geant4.cern.ch]. Application of its additional package, called DNA, allows to simulate the behavior of secondary electrons (i.e. to trace their trajectories) down to energies of about several eV.

To simulate the space distributions of secondary electrons in the Auger blob we have placed an isotropic (spherically symmetrical) source of fast electrons in the origin of coordinates. The initial energies of the emitted electrons were chosen to be 0.5, 1 and 2 keV. GEANT4 takes into account scattering of the electrons and their energy loss processes, but does not take into account any process like ion-electron recombination, electron-electron Coulombic repulsion, trapping and so on. The resulting spatial distributions are shown in Figure 1. (in black, blue and brown). It is seen that at distances larger then 50, 100 and 200 nm the distributions abruptly cut off.

For obtaining the resulting spatial distribution of secondary electrons, we have to average these distributions over the initial energies of the Auger electrons. However, the respective distribution function is not known. So, for simplicity, when calculating the kinetics of local heating of the Auger blob, we approximated $c(r)$ by the Gaussian function (1) with the parameter a equal to 80-100 Å.

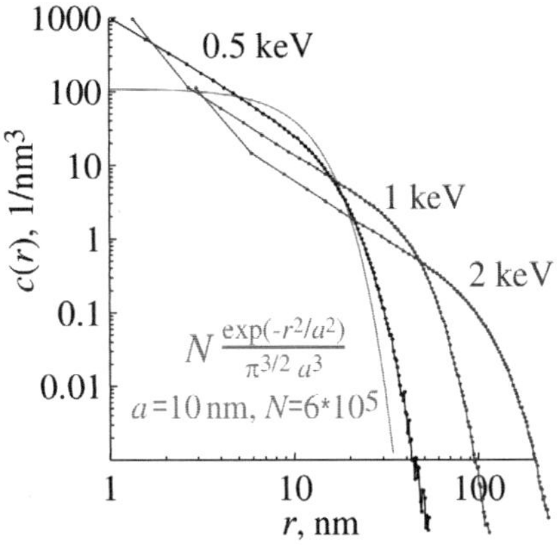

Figure 1. Results of the simulation of the radial distributions, $c(r)$, of the secondary electrons in the Auger blob by means of the GEANT4-DNA toolkit for three different initial energies of the emitted Auger electrons, 0.5, 1 and 2 keV. All distributions are normalized to the total number N of the secondary electrons: $\int c(r)\, 4\pi r^2 dr = N$. The total initial energy of all emitted Auger electrons in each energy group (0.5, 1, 2 keV) was about the same, 11 MeV. It was assumed that the excited iron ion is located at the origin of coordinates and isotropically emits Auger electrons. The red line is an approximation of $c(r)$ by the Gaussian function.

To obtain the distribution of the secondary electrons in the positron blob, we have used monodirectional (along z axis) emission of the positrons with the initial energy $W_{blob} \approx 500$ eV. The resulting axially symmetric distribution is shown in Figure 2.

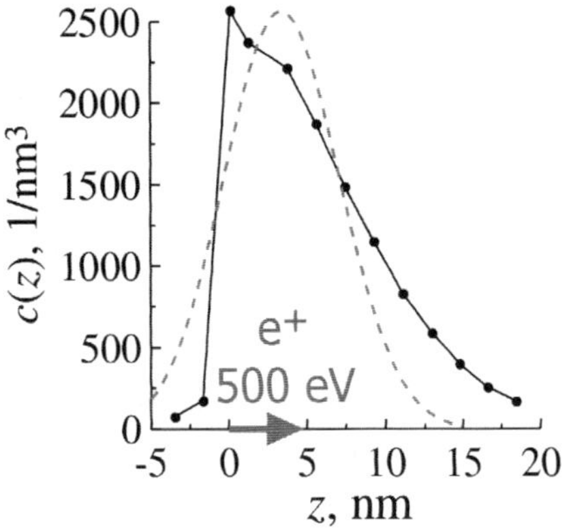

Figure 2. The axially symmetric distribution $c(z) = \int c(\mathbf{r}) dxdy$ of the secondary e- in the positron blob, where e+ loses its last 500 eV of energy. $\int c(z)\, dz = N$ is the total number of the secondary e-. Positrons with energies $W_{blob} = 500$ eV are emitted from the point z=0 along z-axis. In this simulation the total number of all emitted positrons was about 1000. The dashed line is the approximation of $c(z)$ by the Gaussian distribution, shifted toward the direction of the initial momentum of the positron: $c(z) = N \exp(-(z-b)^2/a^2) / \pi^{1/2} a$, where $N \approx 24000$, $a = 5.2$ nm, $b = 3.4$ nm.

3. Effect of local heating in the positron blob and the Auger blob

Assuming that the distributions of secondary electrons in the positron blob and in the Auger blob directly correlate with the distributions of the released energy in space, we can estimate the effect of local heating initiated by the formation of the blobs. The distribution of the local temperature $T(r,t)$ can be obtained by solving the heat transfer equation [8]:

$$c_p\, \rho\, \partial T(r,t)/\partial t = \mathrm{div}\,(\lambda\ \mathrm{grad}\ T(r,t)) + q_+(r,t), \qquad T(r,t=0) = T_{bulk}. \tag{4}$$

Here $T(r,t)$ is the local temperature, T_{bulk} is the temperature of the medium far from the blob, λ is the specific heat conductivity, ρ is the density, $c_p(T(r,t))$ is the heat capacity, which depends on the local temperature $T(r,t)$ of the medium (it is equal to the heat capacity of the solid phase $c_p^S(T)$, if $T(r,t) < T_m$, and to the heat capacity of the liquid, $c_p^L(T)$, if $T(r,t) > T_m$). Finally, $q_+(r,t)$ describes the heat source, i.e.

transformation of the ionization energy losses of the energized particles (positron, Auger electrons, secondary electrons) into heat:

$$q_+(r,t) = W_{blob}\, f(t,\tau)\; c(r,t{=}0)/N\,, \qquad (5)$$

where W_{blob} is the blob formation energy, the function $f(t,\tau) = \exp[-(t\text{-}3\text{ ps})^2/2\tau^2]/(2\pi)^{1/2}\tau$, $\tau = 1$ ps describes the transformation of the energy of primary and secondary particles (into heat) in time and $c(r,t{=}0)/N$ represents the spatial distribution of the released energy.

To utilize Eq. 4 at temperatures close to the melting point T_m of the frozen phase, one may use the following approach [9]. At temperatures slightly below T_m we add to the heat capacity $c_p^S(T)$ of the solid phase a quasi-singular contribution, which takes into account absorption of the enthalpy of fusion. In view of a numerical solution of the heat transfer equation we approximate this contribution by a narrow Gaussian function with the width $\Delta T = 0.25$ °C centered at T_m $-3\Delta T$ (slightly below T_m), namely:

$$c_p(T(r,t)) \;\rightarrow\; c_p(T(r,t)) + q_m \exp[-(T(r,t) - T_m - 3\Delta T)^2/2\Delta T^2]\;/\;(2\pi)^{1/2}\Delta T\,. \qquad (6)$$

An integral over temperature from this additional contribution (the last term in Eq. 6) is equal to the specific latent heat of melting (or the enthalpy of fusion), which is q_m:

$$q_m \int dT\; \exp[-(T(r,t) - T_m - 3\Delta T)^2/2\Delta T^2]\;/\;(2\pi)^{1/2}\Delta T \;=\; q_m\,. \qquad (7)$$

For description of the temperature dependences of the thermal conductivity $\lambda(T)$ and density $\rho(T)$ close to the melting point we have used slightly smoothed (within the interval ~ΔT for convenience of the numerical calculations) step-like approximations, which match values of λ and ρ in the solid and liquid phases: $\lambda(T(r,t){<}T_m) = \lambda^S$, $\lambda(T(r,t){>}T_m) = \lambda^L$ and $\rho(T(r,t){<}T_m) = \rho^S$, $\rho(T(r,t){>}T_m) = \rho^L$.

In the calculations we used the following parameters [10, 11]:

	T_m, [K]	λ^S ; λ^L, [W/m/K]	q_m, [J/g]	c_p^S ; c_p^L, [J/g/K]	ρ^S ; ρ^L, [g/cm^3]	M, [g]
ice / water	273.15	2.38 ; 0.56	334	2.06; 4.18	0.92 ; 1.00	18

In numerical solution of Equation 4 , were designed kinetics of heating / cooling of the blob as a liquid or in a frozen aqueous medium. For the e+ blob we used W_{blob} = 500 eV and a = 40-50 Å, and for the Auger blob W_{blob} = 6 keV and a = 80-100 Å. The calculation results are presented in Figure 3.

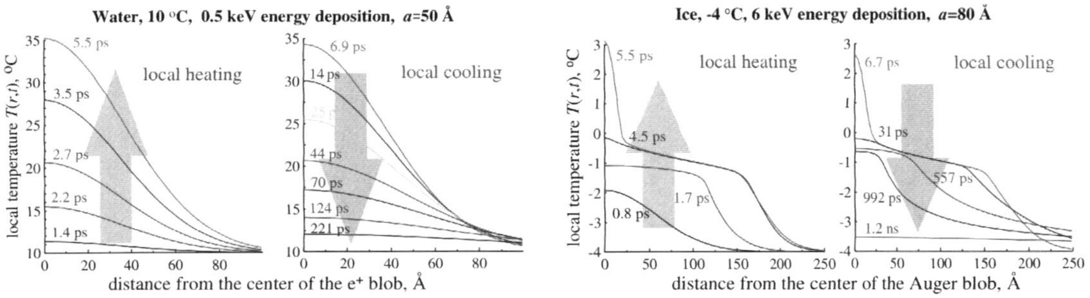

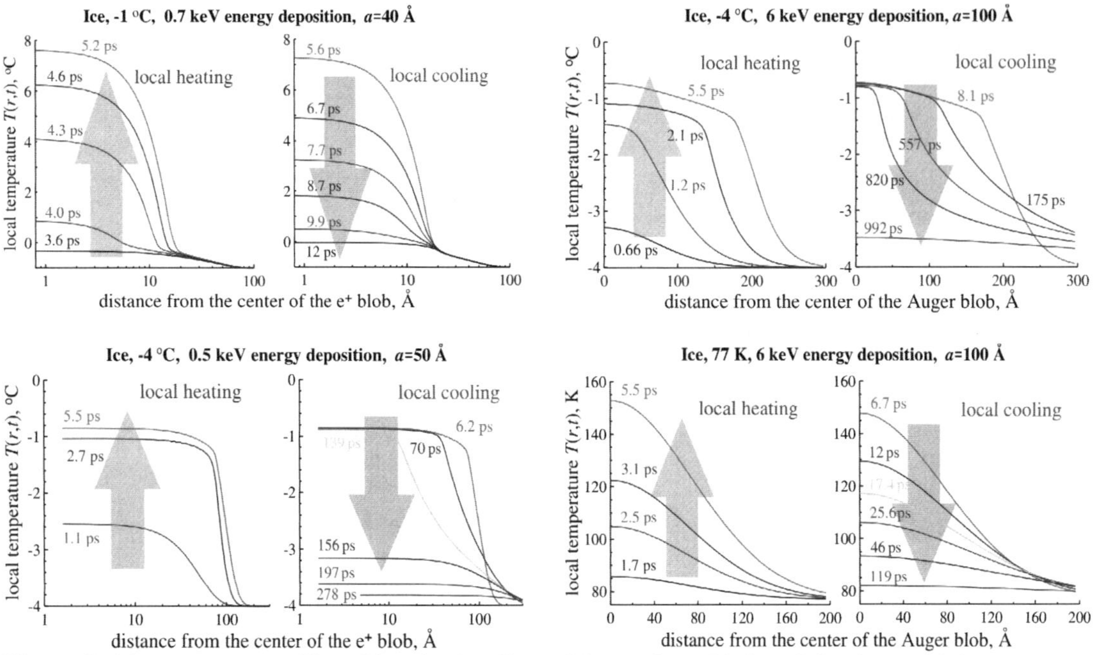

Figure 3. Temperature profiles of heating/cooling of the positron blob and the Auger blob.

4. Conclusion

It is known that during irradiation at low temperatures many intermediate radiolysis products become stabilized. Blob electrons are captured by structural traps, thereby forming trapped electrons, e^-_{tr}. At elevated temperatures, different types of motion become possible and trapped electrons may recombine with counterions and radicals, which results in luminescence (thermoluminescence). It is known that the maximum of thermoluminescence for ice irradiated by gamma-rays at 77 K is observed at $T\approx$110-113 K [12]. The above calculations show that such an increase of local temperature in the Auger blob is quite possible. It may increase the reactivity of e^-_{tr} with respect to intrablob radiolytic products and $^{57}Fe^{3+}$ ions as well. Moreover in EMS experiments at T slightly below T_m , melting of the central part of the Auger blob is possible (for several tens of picoseconds) which may lead to the complete disappearance of the Mössbauer effect.

In the positron blob melting is also possible. We have to note that water is somewhat an inconvenient medium for studying premelting: the thermal conductivity of ice is 10 times larger than that of most molecular media, and the hidden heat of melting is also several times as large. Nevertheless, premelting was observed in ice [4]. So other molecular media could prove to be more convenient for studying premelting effect.

References

[1] Jean Y C, Mallon P E, Schrader D M (Eds.) 2003 *Positron and Positronium Chemistry* (World Scientific, Singapore)
[2] Stepanov S V, Byakov V M, Zvezhinskiy D S, Duplâtre G, Nurmukhametov R R, Stepanov P S 2012 *Adv. Phys. Chem.* **2012** 431962
[3] Zvezhinskiy D S, Stepanov S V, Byakov V M, Zgardzinska B 2013 *Mat. Sci. Forum* **733** 15
[4] Zgardzinska B, Goworek T 2014 *Phys. Lett.* A **378**, 915
[5] Byakov V M, Perfil'ev Yu D, Kulikov L A, Stepanov S V 2009 *Moscow University Chemistry Bulletin* **64** 259
[6] Byakov V M, Kulikov L A, Perfil'ev Yu D, Stepanov S V 2005 *Acta Phys. Pol.* A **107** 792

[7] Stepanov S V, Perfil'ev Yu D, Byakov V M, Kulikov L A 2014 *Acta Phys. Pol.* A **125** 843
[8] Muhieddine M, Canot E, March R 2009 *Int. J. of Fin.* **6** 1
[9] Bonacina C, Comini G, Fasano A, Primicerio M 1973 *Int. J. Heat Mass Tran.* **16** 1825
[10] Yaws C 1999 *Lange's Handbook Of Chemistry*, 15-th Edition (McGRAW-HILL, INC)
[11] The 84th Edition of the Handbook of Chemistry and Physics. CRC Press LLC, 2004
[12] Pikaev A K *Modern radiation chemistry. Solid state and polymers. Applyed aspects* (Moscow, Nauka, 1987)

Thermal annealing of C ion irradiation defects in nuclear graphite studied by positron annihilation

C Q Shi[1], H Schut[2] and Z C Li[1,3,4]

[1] State Key Lab of New Ceramic and Fine Processing, School of Materials Science & Engineering, Tsinghua University, Beijing, China
[2] Department of Radiation Science and Technology, Delft University of Technology, Mekelweg 15, 2629 JB Delft, the Netherlands
[3] Key Laboratory of Advanced Materials (MOE), School of Materials Science & Engineering, Tsinghua University, Beijing, China

Email: zcli@tsinghua.edu.cn

Abstract. In order to investigate the thermal behaviour of radiation induced point defects in nuclear graphite, ETU10 graphite was implanted with 350 keV C^+ ion to doses of 10^{15} and 10^{16} cm^{-2}. The point defects introduced by the implantation were characterized by Positron Annihilation Doppler Broadening (PADB) and their thermal behaviour was studied during "in situ" annealing at Delft Variable Energy Positron beam (VEP). The annealing was performed for 5 minutes at temperatures ranging from 300 K (as implanted) to 1500 K in steps of 100 K. For both doses, an annealing stage at around 450 K is observed followed by a second stage around 700 K. For the high dose implantation vacancy complexes are found which are stable up to a temperature around 1400K.

1. Introduction

Since the building of the first gas-cooled nuclear reactors, graphite has been classified as one of the most promising moderator candidates because of its low atomic number, high thermal conductivity and good mechanical properties[1]. Apart from the moderator function, this material could also be used as reflector and structural materials in the reactor core, which makes nuclear graphite one of the key nuclear materials in the nuclear industry[2]. As a complex polygranular system, nuclear graphite consists two phases: a filler material and a binder phase. The most common filler material is petroleum coke made by delayed coking process[3]. The typically used binder phase in the manufactory is coal tar pitch. After a final stage of graphitization, a very high chemical purity is reached[4]. However, the service performance of nuclear graphite would be greatly impaired as a result of the collision events with fast neutrons, which will lead to the formation of structural defects such as displaced atoms, vacancies and interstitial loops or vacancy aggregations. In addition, Ewel et al also suggested the formation of metastable interstitial-vacancy pair structure in the irradiated nuclear graphite[5]. The subsequent evolution of these defects will lead to dimensional and physical properties changes and the release of the Wigner energy, which is detrimental to the safety of the reactor[6]. Over the past decades, understanding of these defects and their behaviour have always been a front area of the graphite research field[7]. In particular, the behaviour of radiation induced point defects is of technical and scientific interest. Characterization techniques such as transmission electron microscopy (TEM),

[4] To whom any correspondence should be addressed.

scanning tunnelling microscopy (STM) supported by computer simulations[8-10] have greatly enhanced the knowledge about irradiation induced damage of graphite. However, it is still not quite clear what mechanisms play a role in the defect behaviour at atomic level. In this work the nuclear graphite ETU10 was implanted with 350 keV C^+ ions to simulate the collision events with neutrons.

2. Experimental

The samples were manufactured by IBIDEN. It has an apparent density of 1.75 g/cm^3 with low porosity of 15%, and it has a very low ash content of less than 5 ppm. The thermal conductivity is 104.4 W/mK, and the thermal expansion is 3.8×10^{-6} K^{-1}. Two pieces of nuclear graphite ETU10 (20x20x2mm^3), were implanted with 350 keV C^+ ions to doses of 10^{15} and 10^{16} /cm^2, respectively. During the implantations carried out at the Institute of Semiconductors, Chinese Academy of Science, Beijing, the temperature of the samples was kept below 373K.

The point defects introduced by the implantation and their thermal behavior were characterized by the Positron Annihilation Doppler Broadening (PADB) technique at Delft Variable Energy Positron beam (VEP). By selecting positron implantation energies between 0.1 and 25 keV, the maximum obtained positron mean implantation depth is 4 μm. This is well beyond the 1 μm range of the displacement damage distribution (with a 200 nm wide maximum at about 700 nm depth) predicted by SRIM [11]. The "in situ" annealing was performed for 5 minutes at temperatures ranging from 273 K (as implanted) to 1500 K in steps of 100 K. The PADB experiments were started after the samples had cooled down to RT. The measured Doppler broadening *S*-parameters are calculated as the ratio of the counts registered in a fixed momentum window, with $|p_{//}| < 3.5\times10^{-3}$ m_oc, to the total number of counts in the 511 keV photon peak. Here $p_{//}$ is the momentum of the electrons in the direction of gamma emission, m_o is the electron rest mass and c is the speed of light.

3. Results and Discussion

Figure 1 shows the results of the PADB experiments after stepwise annealing of ETU10 irradiated with 10^{15} C^+/cm^2. For clarity reasons the S value of the as-received ETU10 as reference is shown in Figure 3. It has to be clarified that in the interest of a concise plot, the error bars are only shown in Figure 3 for the reference and the first PADB measurement. Based on the same measurement principle, the lengths of the error bars are expected to be at the similar scale for the rest data in Figures 1 and 3.

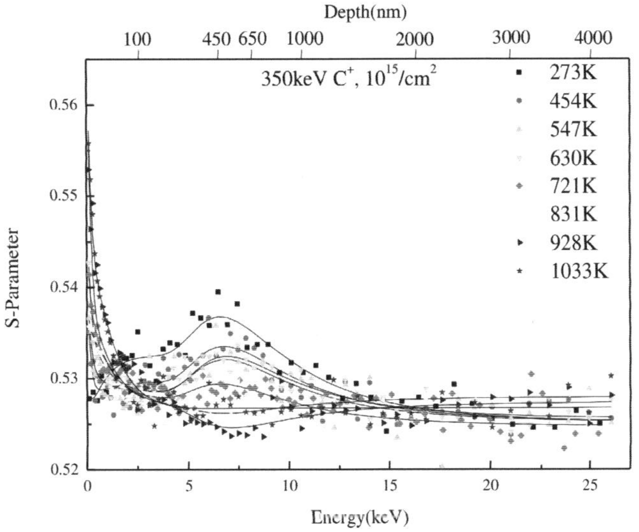

Figure.1 S-Parameter versus positron energy for ETU10 implanted with 10^{15} C^+/cm^2

For the as-received ETU10 specimen, it can be seen that the S values distribute randomly between 0.520 and 0.530, which indicates that positrons show no significant depth preference when annihilate within the penetration range. Therefore, we could take this as the reference when compared with the as-irradiated data. In Figures 1 and 3 the latter is displayed by the scatter symbols, while the full lines through the symbols are the results of VEPFIT[12] calculations using a three-layer-model. The width of the first two layers is derived from the SRIM calculations. The mean implantation depth scale of positron is shown on the top of the chart. The expression of the calculation is described elsewhere[13]. The first layer extends up to 450 nm beneath the implantation surface followed by the second layer with a width of 200 nm. The third and last layer covers the range up to the maximum positron implantation depth and represents the undamaged region. Thus, S_1, S_2, S_3 are the fitted Doppler broadening parameters of the first, second layer and the bulk, respectively. As can be seen, the data of the "as-implanted" sample (273 K) shows a maximum of about 0.54 at the positron implantation energy of about 7 keV. This energy corresponds to a mean positron implantation depth of about 500 nm. This maximum is well above the average value of 0.525 found for the non-implanted region, and therefore is representative for defects introduced by the carbon ion irradiation. With increasing annealing temperature the maximum in the S curve decreases to the level obtained for the reference sample. Figure 2 shows the values of the fitted Doppler parameters (S_1, S_2 and S_3) as a function of annealing temperature.

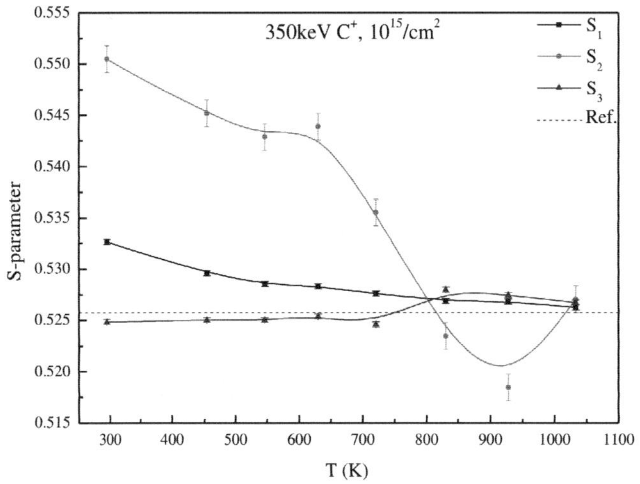

Figure 2. Fitted S values as a function of the annealing temperature. The uncertainties in the fitted parameters are obtained by VEPFIT.

The clear drop of the S_2 value after annealing at 450 K is consistent with the well accepted Wigner energy release at about 473 K[14]. This could be interpreted as the annihilation of the Frenkel pair defects, that is, the vacancies are mobile at 473 K to annihilate adjacent interstitial[15]. At annealing temperatures between 450 K and 600K, the S-parameter decreases slowly while a next sharp decrease is observed at 700 K. As Wigner energy may release at different stages, the significant reduction may result from more released Wigner energy. Accompanied by the energy release, the most possible internal process is the monovancancy migration and in case of high vacancy density aggregation into multi-vacancy complexes, which could only be removed by elevated annealing temperature[15]. A next sharp decrease is observed at annealing temperature higher than 700K where after S_2 decreases quickly to the bulk S value. For the moment (based on the Doppler broadening data only) it is difficult to explain the low S_2 value at the temperature around 900 K. Usually, a low S parameter is associated with small open-volumes, and an S parameter lower than that of the reference would hint at volumes smaller than that associated with interstitial dimensions. Possibly, this could be verified by future positron annihilation lifetime experiments. The Doppler parameter S_1 representing the near surface defects, is less influenced by the implantation and the effect of annealing is observed as a gradual decrease without evidence of annealing stages. Apparently, almost all the defects (monovacancies) have been removed, which shows that defects induced by low dose ion irradiation can be easily removed at lower annealing temperature. Figure 3 presents the PADB measurements of ETU10 irradiated at a ten times higher C ion dose of $10^{16}/cm^2$. The observed higher maximum of the S-parameter indicates that more defects (or larger open volume defects) are introduced. In addition, an increase of S in the first near surface layer is clearly present.

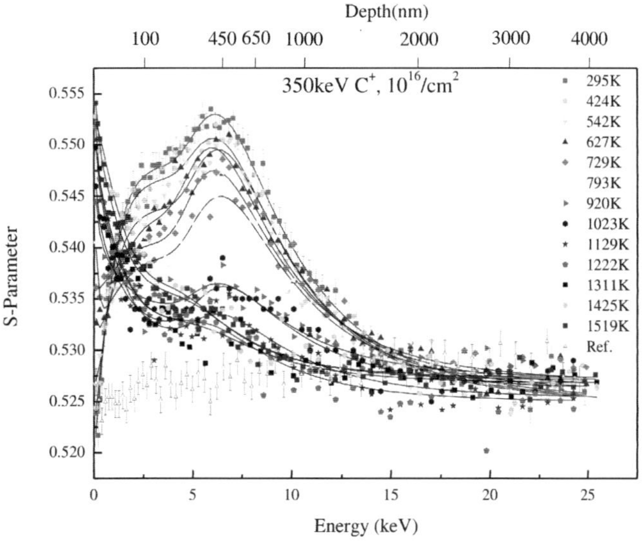

Figure 3. S-Parameter versus positron energy for ETU10 implanted with 10^{16} C^+/cm^2.

Figure 4 shows the values of the fitted Doppler parameters (S_1, S_2 and S_3) as a function of annealing temperature. (Note the extended temperature scale in comparison to figure 2).

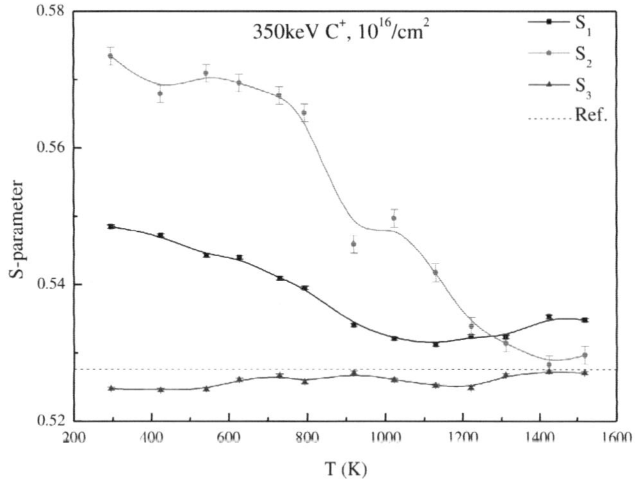

Figure 4. Fitted S values as a function of the annealing temperature.

As in the case of the dose of 10^{15}/cm^2, the first annealing stage in S_2 is seen at a temperature 400 K, which also derives from the characteristic Wigner energy release. With the annealing temperature increasing to 800 K, the S_2 value descends gradually, followed by a sharp decrease between 800K and 920 K. In case of the low dose implantation at this temperature the majority of the small vacancy defects have been annealed out. For the high dose implantation this is clearly not the case and therefore the annealing was continued. At the next annealing temperature of about 1000K first an increase in S_2 is observed. It is noteworthy that a similar increase is seen in the analyses of the low dose implanted sample. In this case the increase started from a lower level associated with the bulk. Further annealing results in a gradual decrease in S towards the bulk value. However, the maximum in the S curve is still higher than that of the reference data. Tang[16] suggests an aggregation structure formed by 6 vacancies as the possible stable cluster in heavily irradiated graphite. The planar V_6 rings may survive even at temperature up to 1773 K as seen in this study.

The annealing behaviour of S_1 for the high dose implantation shows a trend similar to that of S_2 for the low dose in which the effect of vacancy clustering is less pronounced because of the lower initial vacancy concentration. The exception is that in the near surface region the bulk value is not fully reached. A similar effect is seen in our recent studies on 200 keV He irradiated nuclear graphite [17] and may be caused by surface erosion by ion impact.

4. Conclusion

Defects in nuclear graphite ETU10 have been introduced by 350 keV C-ion implantation to doses of 10^{15} and 10^{16} C^+/cm^2, respectively. The thermal evolution of the point defects has been studied by Positron Annihilation Doppler Broadening after “in situ” annealing up to 1500 K. For both doses an annealing stage at around 450 K is observed followed by a second stage around 700 K. For the high dose implantation vacancy complexes are found which are stable up to a temperature around 1400K.

Acknowledgments

The authors are grateful to the Koninklijke Nederlandse Akademie van Wetenschappen (KNAW, the Netherlands), the National Science and Technology Major Project (under Grant 2011ZX06901-017 and 2008ZX06901-001, China), and the National Natural Science Foundation of China (under Grant 61176003, China). We thank Professor Tadashi Maruyama of Tokyo Institute of Technology for providing graphite samples and the necessary information about the samples.

References

[1] Marsden B J, Wickham A J 2006 *IAEA TECDOC*-1521
[2] Chen D Y, Li Z C, Miao W, Zhang Z 2012 *Mater. Trans.* **53** 1159
[3] Burchell T.D. 2012 *Material Properties/Oxide Fuels for Light Water Reactors and Fast Neutron Reactors* (Comprehensive Nuclear Materials vol.2, Amsterdam: Elsevier)
[4] Zhou Z, Bouwman W G, Schut H, Pappas C 2014 *Carbon* **69** 17
[5] Telling R H, Heggie M I 2007 *Philos. Mag.* **87** 4797
[6] Ewels C P, Telling R H, El-Barbary A A, Heggie M I, Briddon P R 2003 *Phys. Rev. Lett.* **91** 025505
[7] Banhart F 1999 *Rep. Prog. Phys.* **62** 1181
[8] Bollmann W 1961 *J. Appl. Phys.* **32** 869
[9] Liu J, Yao H J, Sun Y M, Duan J L, Hou M D, Mo D, Wang Z G, Jin Y F, Abe H, Li Z C, Sekimura N 2006 *Nucl. Instrum. Methods Phys. Res.* B **245** 126
[10] Muto S, Horiuchi S, Tanabe T 1999 *J. Electron Microsc.* **48** 767
[11] Ziegler J F, Biersack J P, Ziegler M D 2008 *SRIM - The Stopping and Range of Ions in Matter* (Chester, Maryland: SRIM Company)
[12] Veen A van, Schut H, Clement M, Nijs J M M de, Kruseman A, IJpma M R, 1995 *Appl. Surf. Sci.* **85** 216

Journal of Physics: Conference Series **674** (2016) 012019 doi:10.1088/1742-6596/674/1/012019

[13] Carvalho I, Schut H, Fedorov A, Luzginova N, Sietsma J 2013 *J. Nucl. Mater.* **442** S48
[14] Mitchel E W J and Taylor M R 1965 *Nature* **208** 638
[15] Latham C D, Heggie M I, Alatalo M, Öberg S, Briddon P R 2013 *J. Phys. Condens. Matter* **25** 135403
[16] Tang Z, Hasegawa M, Shimamura T, Nagai Y, Chiba T, Kawazoe Y, Takenaka M, Kuramoto E, Iwata T 1999 *Phys. Rev. Lett.* **82** 2532
[17] Hu Z, Li Z C, Zhou Z, Shi C Q, Schut H, Katia P 2014 *J. Phys.: Conf. Ser.* **505** 012014

Journal of Physics: Conference Series **674** (2016) 012020 doi:10.1088/1742-6596/674/1/012020

Computational studies of positron states and annihilation parameters in semiconductors – vacancy-type defects in group-III nitrides –

S Ishibashi[1] and A Uedono[2]

[1] Nanosystem Research Institute (NRI) "RICS", National Institute of Advanced Industrial Science and Technology (AIST), Tsukuba, Ibaraki 305-8568, Japan
[2] Division of Applied Physics, Faculty of Pure and Applied Science, University of Tsukuba, Tsukuba, Ibaraki 305-8573, Japan

E-mail: shoji.ishibashi@aist.go.jp

Abstract. We have computationally studied positron sates and annihilation parameters in semiconductors, especially in group-III nitrides. A random alloy system $In_{0.5}Ga_{0.5}N$ was model with the special-quasirandom-structure scheme and distributions of annihilation parameters for cation monovacancies and divacancies were investigated. On GaN, we calculated annihilation parameters considering spin polarization for Ga vacancies with various charge sate and demonstrated how the positron annihilation technique is useful to study defect-induced or mediated magnetism in dilute magnetic semiconductors. We also made calculations based on the two-component density functional theory and compared their results with those obtained by the conventional scheme.

1. Introduction

Group-III nitrides are important and promising materials for various applications such as optoelectronics etc. Their properties and functions can be controlled by forming alloys and/or multilayer structures. For group-III nitride based devices, often, defects such as vacancies and dislocations play a critical role in their performances. The positron annihilation technique is known as a powerful tool to study (vacancy-type) defects in various solids including semiconductors [1,2]. We have been applying the positron annihilation technique in characterizing defects in group-III nitrides [3-5]. In this paper, we would like to report our recent progress on the computational side.

Alloys of group-III nitrides such as (Al,Ga)N and (In,Ga)N are frequently used in actual devices. It is thought that two kinds of cations are randomly distributed in these systems. To model atomic arrangements in (In,Ga)N as a random alloy, we utilized the special-quasirandom-structure (SQS) approach [6,7] and calculated positron states and annihilation parameters in bulk $In_{0.5}Ga_{0.5}N$, $In_{0.5}Ga_{0.5}N$/GaN and V_{cation} therein [5]. In the following, results for cation-nitrogen divacancies are reported.

Diluted magnetic semiconductors have been attracting much attention. Although group-III nitrides themselves are nonmagnetic *sp* systems, they can be modified to be magnetic. Gd-doped GaN is an example. On this system, Dhar *et al.* observed a large magnetic moment of 4000 μ_B per Gd atom [8]. Dev *et al.* investigated cation-vacancy induced magnetism in GaN and BN by means of electronic structure calculations [9]. Their results suggest the existence of defect-induced or mediated collective magnetism. Gohda and Oshiyama found that V_{Ga}s and Gd atoms are ferromagnetically-coupled and that V_{Ga}s contribute significantly to the large magnetic moment [10]. Here, we made spin-polarized calculations and obtained magnetic Doppler broadening spectra for V_{Ga}^{0}, V_{Ga}^{-1} and V_{Ga}^{-2} in GaN. To illustrate the uniqueness of the positron annihilation technique in studying vacancy assisted ferromagnetism, we calculated magnetic Compton scattering spectra also and made a comparison.

The density of positron delocalized in the bulk can be regarded as vanishingly small. For such a case, it is a reasonable assumption that positrons do not affect the electronic structure and the atomic arrangement. In the computational procedure, the electronic structure is first obtained and, with clamping it, the positron state is then calculated. In the following, we describe this procedure as the conventional scheme. As for a positron trapped at a defect, its density is finite. The electronic structure and the atomic arrangement should be affected by the trapped positron. To deal with such a case, the two-component density functional theory (TC-DFT) scheme [11,12] is appropriate. Nevertheless, the conventional scheme is frequently applied even for trapped positrons. One reason is that the TC-DFT scheme requires much more computational time than the conventional scheme. So far, we also applied the conventional scheme to calculate positron states trapped at various vacancy-type defects in group-III nitrides. To evaluate their accuracies, we have made calculations for divacancies in group-III nitrides and several other semiconductors by means of the TC-DFT.

2. Computational details

All the calculations in the present paper have been performed with our computational code QMAS (Quantum MAterials Simulator) [13].

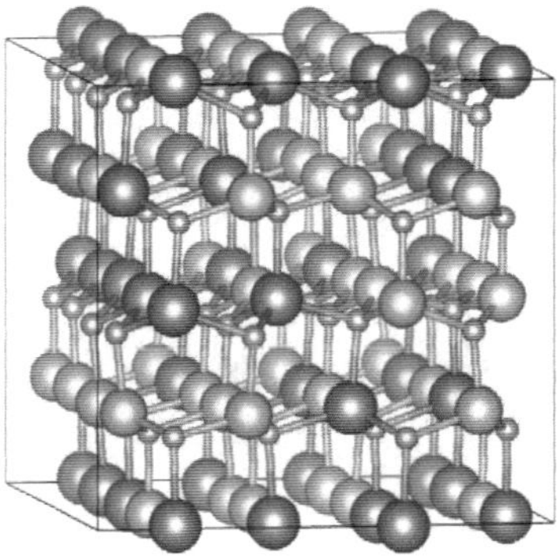

Figure 1. SQS-modelled $In_{0.5}Ga_{0.5}N$ supercell (orthorhombic cell equivalent to 4×4×2 hexagonal unit cell including 128 atoms) used in the calculation. Green, purple and small grey balls represent Ga, In and N atoms, respectively.

2.1. Structural modelling

To calculate positron states and annihilation parameters in group-III nitride systems containing defects or in alloys, we used orthorhombic supercells which are equivalent to 4×4×2 hexagonal unit cells of the wurzite structure. For cubic semiconductors, we used 2×2×2 and 3×3×3 supercells. For $In_{0.5}Ga_{0.5}N$, in which two kinds of cations are randomly distributed, we used the special-quasirandom-structure (SQS) approach [6,7]. This is a method to provide realistic structural modeling for random alloys using periodic supercells. With this method, two kinds of atoms A and B are arranged so that pair-correlation functions become as close to those for an infinite random alloy as possible. Practically, an atomic arrangement, which gives the lowest Warren-Cowley parameter for several neighboring shells, is chosen. The Warren-Cowley parameter for the j-th nearest neighbor shell is expressed as

$$\alpha_j = 1 - \frac{P_B(j)}{x_B}, \qquad (1)$$

where $P_B(j)$ is the probability of finding a B atom on the j-th nearest neighbor shell about an A atom as the origin. In the present study, shells up to the 7-th are taken into account and the following quantity should be minimized:

$$T = \sum_{j=1}^{7} \alpha_j^2 . \tag{2}$$

Although the supercell dimensions are also optimized in the original SQS approach, we fixed them in the present work. In figure 1, the SQS-modelled $In_{0.5}Ga_{0.5}N$ supercell is shown. From this structure, an initial structure for a defect system was prepared by removing (an) atom(s).

2.2. *Electronic structure calculation*

Electronic structures were calculated based on the DFT within the generalized gradient approximation using the exchange-correlation functional of Perdew, Burke and Ernzerhof (PBE) [14]. Electronic wave functions are described with the plane-wave basis in the framework of the projector augmented-wave (PAW) method [15]. The PAW method is adequate to obtain the momentum density distribution of annihilation quanta accurately even in a higher momentum region [16,17]. It is also known to be more efficient than the norm-conserving pseudopotential method. The plane-wave energy cutoff and the force convergence criterion were set to be 20 Ha and 5×10^{-5} Ha/bohr, respectively, for all the calculations with the conventional scheme. As for $\boldsymbol{k}$ point sampling, only the Γ point was used except for spin-polarized calculations on the GaN supercell, where 2×2×2 $\boldsymbol{k}$ points were used.

2.3. *Positron state and annihilation parameters*

The positron effective potential was constructed using core electron wave functions from isolated atom calculations and valence electron wave functions described in the PAW formalism [15]. As for the positron-electron correlation potential term, we used that proposed by Boroński and Nieminen [11]. The positron wave function is also described with the plane-wave basis and optimized in the reciprocal space iteratively. With the obtained positron and electron wave functions, the momentum distribution of annihilation quanta $\rho(\mathbf{p})$ and positron lifetimes τ were calculated with the following equations:

$$\rho(\mathbf{p}) \propto \sum_j \left| \int \psi_+(\mathbf{r}) \psi_{-j}(\mathbf{r}) \sqrt{\gamma(n_-(\mathbf{r}))} \exp(-i\mathbf{p}\cdot\mathbf{r}) d\mathbf{r} \right|^2 , \tag{3}$$

$$1/\tau = \lambda = \pi r_0 c \int n_+(\mathbf{r}) n_-(\mathbf{r}) \gamma(n_-(\mathbf{r})) d\mathbf{r} , \tag{4}$$

where ψ_+, ψ_{-j} , n_+ and n_- are the positron wave function, the j-th electron wave function, the positron density and the electron density, respectively. r_0 represents the classical electron radius and c is the speed of light. γ is the enhancement factor. We used the Boroński-Nieminen enhancement factor

$$\gamma_{\mathrm{LDA}} = 1 + 1.23 r_s + 0.8295 r_s \sqrt{r_s} - 1.26 r_s^2 + 0.3286 r_s^2 \sqrt{r_s} + r_s^3/6 , \tag{5}$$

where $r_s = [3/(4\pi n_-)]^{1/3}$ is the density parameter, with a small modification to deal with semiconductors [18]. Here, in analogy with the GGA enhancement factor [19], and to be consistent with the positron-electron correlation energy, we used the expression

$$\gamma_{\mathrm{semi}} = 1 + (\gamma - 1)(1 - 1/\varepsilon_\infty) . \tag{6}$$

ε_∞ is the dielectric constant of the target system. Although it can be determined by first-principles calculations, we adopted the experimental values for convenience. In case of alloys, linearly-interpolated values were used. For reference, the expression in the original semiconductor model [18] is

$$\gamma_{\text{semi}} = 1 + \frac{(1-1/\varepsilon_\infty)r_s^3 + 10}{6}. \tag{7}$$

When we make comparisons with results obtained from TC-DFT calculations, we used

$$\gamma_{\text{semi}} = 1 + 1.23r_s + 0.9889r_s\sqrt{r_s} - 1.4820r_s^2 + 0.3956r_s^2\sqrt{r_s} + (1-1/\varepsilon_\infty)r_s^3/6, \tag{8}$$

which is compatible with that used in the TC-DFT calculations. The enhancement factor expressed in (8) corresponds to that in the vanishing n_+ limit for the TC-DFT scheme. Doppler broadening spectra were calculated by integrating equation (3)

$$J(p_z) = \iint \rho(\mathbf{p}) dp_x dp_y. \tag{9}$$

If spin sign is distinguished, it is expressed as

$$J(p_z) = \iint (\rho_{\text{up}}(\mathbf{p}) + \rho_{\text{down}}(\mathbf{p})) dp_x dp_y. \tag{10}$$

From magnetic Doppler broadening measurements,

$$J_{\text{mag}}(p_z) = \iint (\rho_{\text{up}}(\mathbf{p}) - \rho_{\text{down}}(\mathbf{p})) dp_x dp_y. \tag{11}$$

Then, we can obtain

$$J_{\text{up}}(p_z) = \frac{1}{2}(J(p_z) + J_{\text{mag}}(p_z)) = \iint \rho_{\text{up}}(\mathbf{p}) dp_x dp_y, \tag{12}$$

$$J_{\text{down}}(p_z) = \frac{1}{2}(J(p_z) - J_{\text{mag}}(p_z)) = \iint \rho_{\text{down}}(\mathbf{p}) dp_x dp_y. \tag{13}$$

To calculate Compton scattering spectra, the following expression is used instead of equation (3):

$$\rho(\mathbf{p}) \propto \sum_j \left| \int \psi_{-j}(\mathbf{r}) \exp(-i\mathbf{p}\cdot\mathbf{r}) d\mathbf{r} \right|^2. \tag{14}$$

2.4. Two-component density functional (TC-DFT) scheme

According to [12], we made calculations with following steps:
(1) The electronic structure is calculated neglecting the presence of positron.
(2) The positron state is calculated with the results obtained in the (1) or (3) step.
(3) The electronic structure is calculated considering the positron state obtained in (2).
(4) Steps (2) and (3) are repeated to obtain the convergence of the electron and positron states.
(5) The atomic positrons are optimized calculating forces considering the positon effect.

Steps (1)-(5) are repeated to obtain the converged result. Because of steps (2)-(4), more computational time is required compared with the conventional method. The force convergence criterion was set to 5×10^{-5} Ha/bohr.

As for the electron-positron correlation energy and potentials, we used the expressions and interpolation scheme in [12]. Correspondingly, we used the enhancement factor evaluated by the interpolation scheme in [11,12] with a modification for semiconductors [20].

3. Results and discussion

Measured Doppler broadening spectra are often analysed through *S* and *W* parameters. As shown in figure 2, the *S* parameter is defined as the ratio of the center area to the total spectrum area while the *W* parameter is that for the higher momentum areas. Positrons trapped at a defect usually give a higher *S* value and a lower *W* value since an increase of the open volume size results in narrowing of the spectrum. For all the calculations in the present study, S is evaluated in the momentum region between ± 0.419 a.u. while W is evaluated in the regions of ± (1.828~3.688) a.u. Plotting S and W together on a two-dimensional map is useful to identify defect species. If there is only one type of defect in the system, observed parameters S_{obs} and W_{obs} should be described with the parameters for the bulk S_b and W_b and those for the defect S_d and W_d as $S_{obs} = f_bS_b + f_dS_d$ and $W_{obs} = f_bW_b + f_dW_d$ ($f_b + f_d = 1$). The point (S_{obs}, W_{obs}) should be located on the line connecting (S_b, W_b) and (S_d, W_d) shown in figure 3(a). If there are two types of defects, denoting parameters for the second defect as $S_{d'}$ and $W_{d'}$, $S_{obs} = f_bS_b + f_dS_d + f_{d'}S_{d'}$ and $W_{obs} = f_bW_b + f_dW_d + f_{d'}W_{d'}$ ($f_b + f_d + f_{d'} = 1$). In this case, the point (S_{obs}, W_{obs}) should be located inside the triangle connecting (S_b, W_b), (S_d, W_d) and ($S_{d'}$, $W_{d'}$) shown in figure 3(b). In practical situations, the point (S_{obs}, W_{obs}) moves by, for example, annealing. In a certain case, vacancies aggregate to form vacancy clusters, while, in another case, vacancies form complexes with impurity atoms. Theoretical predictions are expected to be useful to distinguish them.

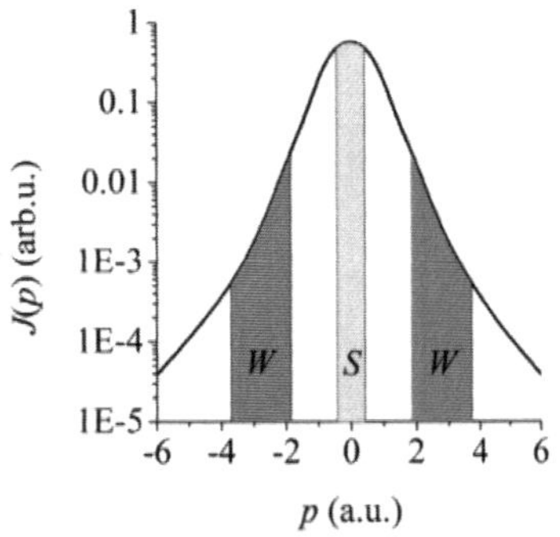

Figure 2. Definition of S and W. S and W are given as ratios of the shaded areas to the total area.

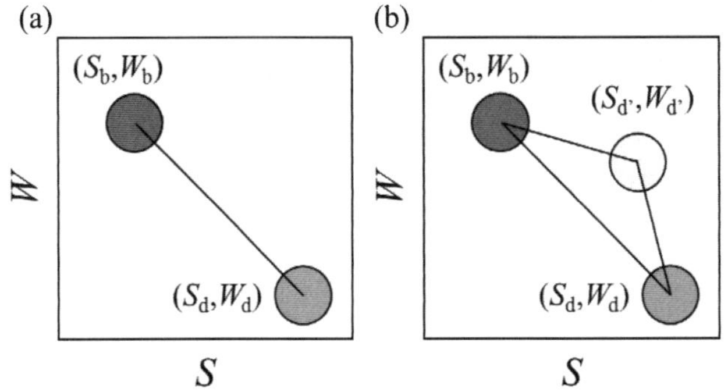

Figure 3. Schematic S–W plots for the positron annihilation from the defect-free state (S_b, W_b) and the trapped state by defects which show different (S,W) values: [(S_d, W_d) and ($S_{d'}$,$W_{d'}$)].

3.1. Divacancy in ($In_{0.5}Ga_{0.5}$)N

In our previous work [5], we calculated S and W parameters for 64 types of cation vacancies in ($In_{0.5}Ga_{0.5}$)N (a supercell modelled by the SQS approach [6,7]) and found an extended distribution of S in the S–W plot. Tuomisto *et al.* also reported similar finite spreads of positron annihilation parameters (S, W and the positron lifetime) for cation monovacancies in (In,Ga)N systems [21].

In the present work, we calculated S and W parameters for 104 types of cation-nitrogen divacancies, which are randomly selected among 256 possibilities, in ($In_{0.5}Ga_{0.5}$)N. The results are plotted together with those for the bulk and cation monovacancies in figure 4. As for the S parameter, significant variation has been observed and there is a finite overlap in the S–W plot between two types of defects. We also plot each S, W and the positron lifetime τ in figure 5 as a function of x_{Ga}, which represents the gallium concentration of the neighboring cation shells containing 12 sites. Although it is somewhat obscure, there seems to be positive correlation.

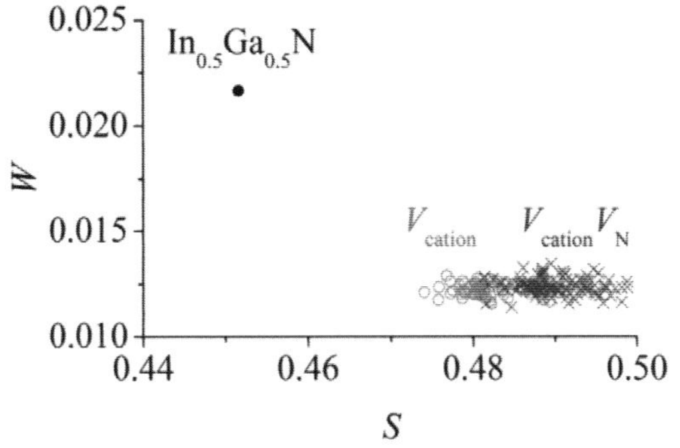

Figure 4. *S–W* plot for bulk $In_{0.5}Ga_{0.5}N$, cation monovacancies and cation-nitrogen divacancies.

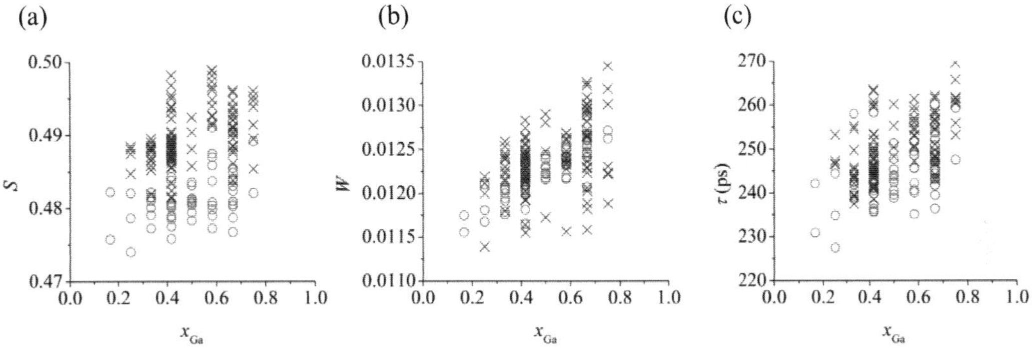

Figure 5. Positron annihilation parameters (a) *S*, (b) *W*, (c) *τ* for cation vacancies (○) and cation-nitrogen divacancies (×) as a function of neighbouring gallium concentration x_{Ga}.

3.2. Vacancy assisted ferromagnetism in GaN

First, spin-polarized calculations were made on the 4×4×2 supercell containing a gallium monovacancy ($Ga_{63}N_{64}$) with optimizing the atomic positions. In figure 6, obtained spin density distributions around V_{Ga}^{0}, V_{Ga}^{1} and V_{Ga}^{2} are shown. Dumbbell-like spin density distributions are observed. Each of them corresponds to the 2*p*-like orbital of nitrogen. The gallium vacancy is surrounded by four major dumbbells. The calculated net magnetic moment values are 3 μ_B, 2 μ_B and 1 μ_B for V_{Ga}^{0}, V_{Ga}^{-1} and V_{Ga}^{-2}, respectively.

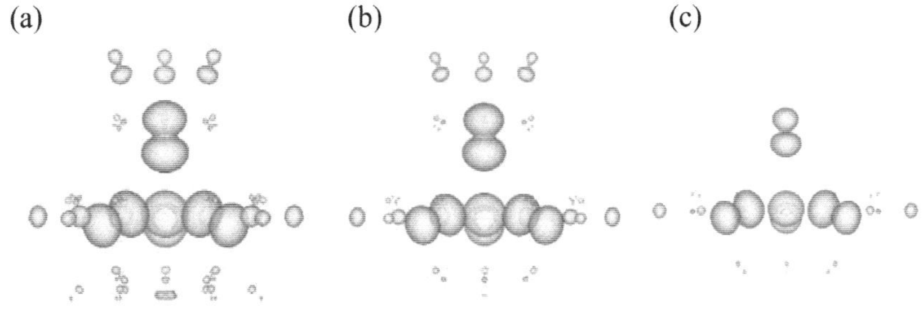

Figure 6. Spin density distribution around (a) V_{Ga}^{0}, (b) V_{Ga}^{-1} and (c) V_{Ga}^{-2}. Each dumbbell-like distribution corresponds to the 2*p*-like orbital of nitrogen. The gallium vacancy is surrounded by four major dumbbells.

Magnetic Doppler broadening and Compton scattering spectra were calculated along the [1$\bar{1}$00] direction for three systems mentioned above. The obtained results are shown as $(\rho_{up}-\rho_{down})/(\rho_{up}+\rho_{down})$

Journal of Physics: Conference Series **674** (2016) 012020 doi:10.1088/1742-6596/674/1/012020

in figure 7. It is clearly shown that the magnetic Doppler spectra are approximately 20 times enhanced than the magnetic Compton ones. This is because positrons are selectively trapped at V_{Ga}. The positron annihilation Doppler broadening technique is expected to be uniquely sensitive for vacancy-assisted ferromagnetism. It is also noted that curve intervals are more or less similar to each other reflecting the magnetic-moment difference in the graph of the Compton scattering spectra while it is not the case with the Doppler broadening spectra. It should be attributed to the positron density distribution localized at V_{Ga}. A detailed analysis is ongoing. For reference, the (S, W) values were evaluated from the spin-integrated spectra as (0.471, 0.0163), (0.474, 0.0146) and (0.474, 0.0135) for V_{Ga}^{0}, V_{Ga}^{-1} and V_{Ga}^{-2}, respectively. The S parameter is almost unchanged while the W parameter decreases slightly with the charge state varying from neutral to doubly negative.

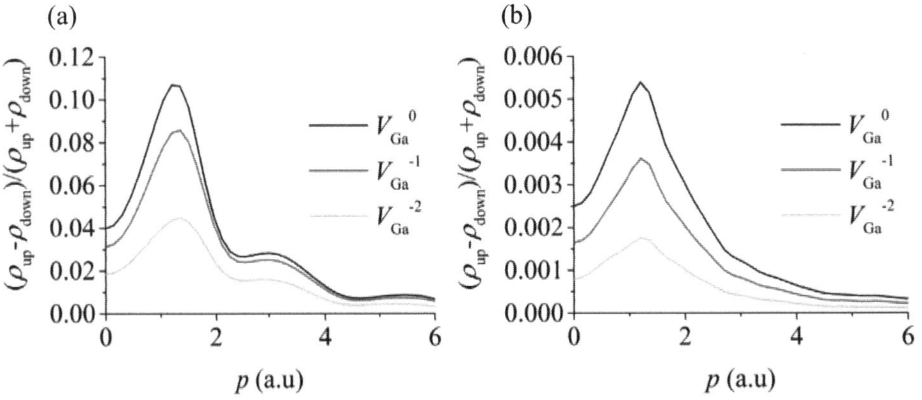

Figure 7. One-dimensional projection of spin density distribution along [1$\bar{1}$00] for V_{Ga}^{0}, V_{Ga}^{-1} and V_{Ga}^{-2}: (a) Doppler broadening and (b) Compton scattering

3.3. Two-component density functional theory scheme

First, we would like to mention our calculations for monovacancy (V) and divacancy (V_2) in Si. The calculated lifetime values are listed in table 1. It is clearly shown that there is a strong supercell-size dependence for the conventional scheme. The main reason is the difference in the atomic arrangement in the vicinity of V or V_2. The larger supercell, the larger inward relaxation is observed. This tendency and the defect structures are consistent with the previous studies [22,23]. On the other hand, for the TC-DFT scheme, the resultant lifetimes are longer than those obtained with the conventional scheme and show a much smaller supercell-size dependence. The monovacancy is surrounded by four Si atoms (1~4) showing the T_d point symmetry in the unrelaxed structure. Six distances between two of them (d_{12},d_{13},d_{14},d_{23},d_{24},d_{34}) are equivalent to each other in this case and the value is 3.84 Å. After structural relaxation, with the 2×2×2 supercell, the distance decreases to 3.67 Å (−4.4 %) for the conventional scheme while it increases to 4.00 (+4.2 %) for the TC-DFT scheme. With the 3×3×3 supercell, the point symmetry becomes D_{2d}. There are four equivalent distances (d_{13},d_{14},d_{23},d_{24}) of 3.45 Å (−10.1 %) and two distances (d_{12}, d_{34}) of 3.00 Å (−21.9 %) for the conventional scheme while four distances of 3.93 Å (+2.4 %) and two distances of 3.70 Å (−3.7 %) for the TC-DFT scheme. The presence of positron suppresses the inward relaxation around V. A similar situation is observed for V_2 as described below together with results of other semiconductors. This has been already found and reported by Saito and Oshiyama [24] although they used a different version of TC-DFT [25]. They obtained 279 ps for V and 309 ps for V_2 with the 2×2×2 supercell. Considering that they did not modify the enhancement factor with ε_∞, their values seem to be slightly overestimated. In other words, the positron effect seems to be slightly overestimated.

Table 1. Calculated positron lifetimes (in ps) for monovacancy (V) and divacancy (V_2) in Si with the conventional scheme and the TC-DFT scheme on different-sized supercells.

	Conventional		TC-DFT	
	2×2×2	3×3×3	2×2×2	3×3×3
V	232.5	222.8	267.7	268.6
V_2	282.7	247.4	314.8	315.2

Table 2. Distance between atoms neighbouring divacancy (in Å). In parentheses, relative changes from the unrelaxed values are shown (in %).

		Conventional		TC-DFT	
		cation	anion	cation	anion
C	2×2×2	2.69 (+6.7)		2.75 (+9.2)	
	3×3×3	2.65 (+5.2)		2.73 (+8.4)	
Si	2×2×2	3.62 (−5.7)		3.85 (+0.2)	
	3×3×3	3.01 (−21.7), 3.04 (−20.9), 3.55 (−7.5)		3.69 (−3.9)	
Ge	2×2×2	3.40 (−14.9)		3.50 (−12.4)	
	3×3×3	3.16 (−20.9)		3.40 (−14.8)	
3*C*-SiC	2×2×2	3.88 (+1.0)	4.15 (+8.1)	3.99 (+3.9)	4.20 (+9.5)
	3×3×3	3.08 (−0.2)	3.29 (+6.8)	3.16 (+2.6)	3.35 (+8.6)
AlN	4×4×2	3.67 (+18.0)	3.36 (+8.0)	3.76 (+20.9)	3.32 (+6.7)
GaN	4×4×2	2.98 (−6.5)	3.39 (+6.4)	3.05 (−4.4)	3.35 (+4.9)
InN	4×4×2	3.43 (−3.1)	3.56 (+0.6)	3.50 (−1.1)	3.47 (−2.1)

Thus, the conventional scheme is thought to be not very accurate at least for V and V_2 in Si, and we made calculations for divacancies in several semiconductors including group-III nitrides (C, Si, Ge, 3*C*-SiC, AlN, GaN and InN). Each anion (cation) vacancy forming a divacancy has three nearest neighbor cation (anion) atoms. For elemental semiconductors, although there is no distinction between cation and anion, there are similar 3×2 neighbor atoms. In table 2, distances between two of three atoms are listed. Except for the case of Si with the 3×3×3 cell and the conventional scheme, three distances are almost equal to each other. Correspondingly, positron annihilation parameters S, W and τ were calculated and the results are plotted in figure 8 as a function of bulk modulus. Roughly speaking, the softer material with the smaller B gives the larger difference between the results of the conventional and TC-DFT schemes. For the cubic materials, it is also confirmed that the supercell-size dependence is larger in softer materials. In figure 8(c), for Si and Ge, it looks that the conventional-scheme results with the smaller (2×2×2) cell are better than those with the larger (3×3×3) cell. As mentioned above, the inward relaxation is suppressed in the smaller cell. This is because the supercell size is too small for complete relaxation. For the TC-DFT case, the inward relaxation is suppressed by the presence of positron. The mechanism is different. The better agreement for the smaller cell is not essential.

For 3*C*-SiC, Wiktor *et al.* reported a TC-DFT lifetime of 195 ps for V_C (217 ps with the different version TC-DFT [25]) and also a value of 154 ps obtained with the conventional scheme [20]. We also made a TC-DFT calculations on for V_C in 3C-SiC and obtained a lifetime of 142 ps. Our value is quite different from the value by Wiktor *et al.* The reason is not clear yet.

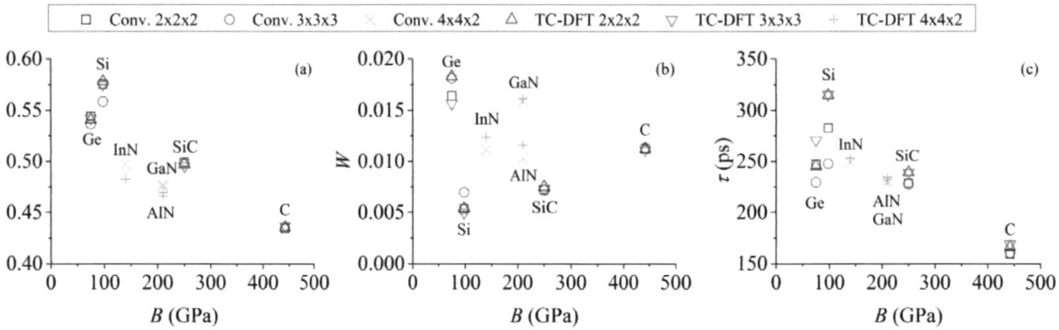

Figure 8. Positron annihilation parameters for the annihilation of positrons trapped by divacancies in various semiconductors. Calculated values of (a) S, (b) W and (c) τ are plotted as a function of bulk modulus.

As shown in figure 8, for C and SiC, the conventional-scheme results are very close to the corresponding TC-DFT ones. On the other hand, there are large deviations for the positron lifetime in Si and Ge between the two schemes. For the other cases, it depends on the situation whether or not the conventional scheme is sufficient to obtain reasonable results. For group-III nitrides, judging from the (S,W) or τ variation shown in figures 4 and 5, the conventional-scheme results can be used for our purpose identifying defect species.

4. Summary

We have made computational studies on defects in semiconductors, especially those in group-III nitrides. Variations of positron annihilation parameters S, W and τ for cation monovacancies and cation-nitrogen divacancies in $(In_{0.5}Ga_{0.5})N$ were examined utilizing the SQS approach to reproduce its random-alloy nature. Magnetic positron annihilation Doppler broadening spectra were calculated for gallium monovancancies with various charge state. The obtained results were compared with magnetic Compton scattering spectra to demonstrate the uniqueness and usefulness of the positron annihilation technique in studying vacancy-assisted ferromagnetism in diluted magnetic semiconductors. For divacancies in several semiconductors, positron annihilation parameters S, W and τ were calculated using the conventional scheme as well as using the TC-DFT scheme. The validity of the conventional scheme has been examined and discussed for various cases.

References

[1] Krause-Rehberg R and Leipner H S 1999 *Positron Annihilation in Semiconductors* (Berlin: Springer)
[2] Tuomisto F and Makkonen I 2013 *Rev. Mod. Phys.* **85** 1583
[3] Uedono A, Ishibashi S, Oshima N and Suzuki R 2013 *Jpn. J. Appl. Phys.* **52** 08JJ02
[4] Uedono A, Sumiya S, Ishibashi S, Oshima N and Suzuki R 2014 *J. Phys.: Conf. Ser.* **505** 012009
[5] Ishibashi S and Uedono A 2014 *J. Phys.: Conf. Ser.* **505** 012010
[6] Zunger A, Wei S-H, Ferreira L G and Bernard J E 1990 *Phys. Rev. Lett.* **65** 353
[7] Mäder K A and Zunger A 1995 *Phy. Rev.* B **51** 10462
[8] Dhar S, Brandt O, Ramsteiner M, Sapega V F and Ploog K H 2005 *Phys. Rev. Lett.* **94** 037205
[9] Dev P, Zue Y and Zhang P 2008 *Phys. Rev. Lett.* **100** 117204
[10] Gohda Y and Oshiyama A 2008 *Phys. Rev.* B **78** 161201(R)
[11] Boroński E and Nieminen R M 1986 *Phys. Rev.* B **34** 3820
[12] Puska M J, Seitsonen A P and Nieminen R M 1995 *Phys. Rev.* B **52** 10947
[13] http://qmas.jp
[14] Perdew J P, Burke K and Ernzerhof M 1996 *Phys. Rev. Lett.* **77** 3865

[15] Blöchl P E 1994 *Phys. Rev.* B **50** 17953
[16] Ishibashi S 2004 *Mater. Sci. Forum* **445-446** 401
[17] Makkonen I, Hakala M and Puska M J 2006 *Phys. Rev.* B **73** 035103
[18] Puska M J, Mäkinen S, Manninen M and Nieminen R M 1989 *Phys. Rev.* B **39** 7666
[19] Barbiellini B, Puska M J, Torsti T and Nieminen R M 1995 *Phys. Rev.* B **51** 7341
[20] Wiktor J, Jomard G, Torrent M and Bertolus M 2013 *Phys. Rev.* B **87** 235207 (Note that the sign of the fourth term in their equation (A9) is wrong.)
[21] Tuomisto F, Norrman V and Makkonen I 2014 *J. Phys.: Conf. Ser.* **505** 012042
[22] Puska M J, Pöykkö S, Pesola M, and Nieminen R M 1998 *Phys. Rev.* B **58** 1318
[23] Iwata J, Shiraishi K and Oshiyama A 2008 *Phys. Rev.* B **77** 115208
[24] Saito M and Oshiyama A 1996 *Phys. Rev.* B **53** 7810
[25] Gilgien L, Galli G, Gygi C and Car R 1994 *Phys. Rev. Lett.* **72** 3214

Theoretical calculation of positron affinities of solute clusters in aluminum alloys

Masataka Mizuno [1, 2], Hideki Araki [1, 2] and Yasuharu Shirai [3]

1 Center for Atomic and Molecular Technologies, Graduate School of Engineering, Osaka University, 2-1 Yamadaoka, Suita, Osaka 565-0871, Japan
2 Division of Materials and Manufacturing Science, Graduate School of Engineering, Osaka University, 2-1 Yamadaoka, Suita, Osaka 565-0871, Japan
3 Department of Materials Science and Engineering, Graduate School of Engineering, Kyoto University, Sakyo, Kyoto 606-8501, Japan

mizuno@mat.eng.osaka-u.ac.jp

Abstract. We have performed theoretical calculations of positron states for solute clusters in aluminum alloys to estimate the positron affinity of solute clusters. Positron states of solute clusters in aluminum alloys were calculated under the electronic structures obtained by first-principles molecular orbital calculations using Al158-X13 clusters. We defined the positron affinity of the solute clusters by the difference in the lowest potential sensed by positrons between the solute clusters and Al bulk. With increasing atomic number of 3d metals, the annihilation fraction of the solute clusters rapidly increases at Mn and shows a maximum at Ni. A similar trend is observed for 4d metals. The localization of positron at the solute clusters mainly arises from charge transfer from Al matrix to solute clusters. The positron affinity defined in this work well represents the localization of positron at the solute clusters in aluminum alloys.

1. Introduction

It is well known that positron annihilation spectroscopy is a very sensitive tool for vacancy-type defects in solids. However, if solute atoms have a stronger affinity to positrons than the matrix atoms, the positron wave-functions are localized at the solute clusters. Nagai *et al.* have investigated ultrafine Cu precipitates in a very dilute Fe-Cu alloy using coincidence Doppler broadening (CDB) method and have found that the wave functions of all positrons are entirely confined within the nanosize Cu precipitates [1]. They also reported that positrons are trapped at solute clusters in Al-Zn and Al-Ag alloys [2]. On the other hand, Puska *et al.* reported calculated positron affinities of elemental metals [3]. Although the positron affinities of elemental metals were calculated using their own structures, the calculated positron affinities agree well with the experimental results reported by Nagai *et al.*: the solute atoms have a stronger positron affinity than the matrix atoms in Fe-Cu, Al-Zn and Al-Ag alloys. Bharathi *et al.* have calculated positron annihilation characteristics of Zn clusters in Al matrix [4]. It was found that positrons form a bound state for Zn clusters and the positron lifetime decreases with increasing Zn cluster size, which agrees with the positron affinities calculated by Puska *et al.* However, it is not clear that the positron affinities of elemental metals are applicable to any alloys. In the present work, we have calculated positron states of solute clusters in aluminum alloys to estimate positron affinities of solute clusters.

2. Computational method

In order to obtain the electron density and potential for calculations of positron wave-functions, the discrete-variational (DV) Xα cluster method using a program code SCAT [5, 6] was employed. We used the model clusters composed of 171 atoms as an aluminum matrix. The orbital populations were calculated based on the Mulliken population analysis [7].

The three dimensional potential $V_+(r)$ sensed by the positron is constructed as a sum of the electron static potential $V_c(r)$ due to the nuclei and the electrons, and a correlation potential $V_{corr}(n_-(r))$ describing the electron-positron correlation:

$$V_+(r) = V_c(r) + V_{corr}(n_-(r)), \quad (1)$$

where $n_-(r)$ is the electron density. For the density dependence of $V_{corr}(n_-(r))$, we used the interpolation formula of Boroński and Nieminen [8]. The Schrödinger equation for the positron was solved by the finite-difference method [9]. The positron lifetime τ is calculated as the reciprocal value of the positron annihilation rate λ by

$$\lambda = \pi r_0^2 c \int n_+(r) n_-(r) \gamma(r) dr, \quad (2)$$

where r_0 is the classical electron radius, c is the speed of the light, $n_+(r)$ is the positron density and $\gamma(r)$ is the enhancement factor describing the short-range pileup of the electron at the positron. For the enhancement factor, we used the interpolation formula [8]. Annihilation fractions of solute clusters were obtained using the Mulliken population analysis.

3. Results and discussion

3.1. Difference in positron lifetimes between solute clusters and solid solutions in Al-Zn alloys

To confirm the localization of positrons at solute clusters, positron lifetimes of Zn clusters were compared to those of solid solution models in which Al and Zn atoms are randomly distributed. The arrangement of Zn atoms was determined so that the pair-correlation functions were close to those in perfect random alloys up to the fifth neighbor shell [10]. Figure 1 shows calculated positron lifetimes of Zn clusters and solid solutions in the Al-Zn alloys as a function of the number of Zn atoms in $Al_{171-x}Zn_x$. Structural relaxations were not considered in either the Zn clusters or solid solution models.

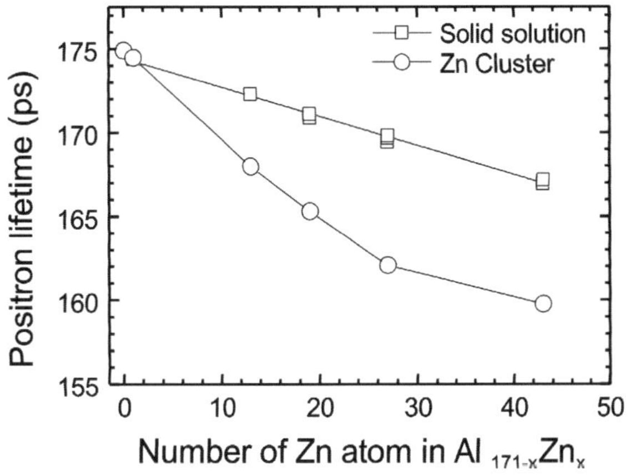

Figure 1. Positron lifetimes of Zn clusters and solid solutions in Al-Zn alloys

Journal of Physics: Conference Series **674** (2016) 012021 doi:10.1088/1742-6596/674/1/012021

The positron lifetimes of the Zn clusters are shorter than those of solid solution models. This result indicates that positrons are localized at the Zn clusters, which results in the shorter positron lifetimes of the Zn clusters. Even in the smallest Zn cluster, Zn13, there is a significant difference between the Zn cluster and solid solution. Therefore, the solute clusters composed of 13 atoms were employed for positron state calculations in the present work.

3.2. Annihilation fractions of clusters composed of 13 solute atoms in Al alloys

The annihilation fraction of solute clusters directly indicates the degree of localization of positrons at solute clusters. Figure 2 shows calculated annihilation fractions of solute atom 13 clusters in the Al alloys. With increasing atomic number of 3d metals, the annihilation fraction of the solute clusters rapidly increases at Mn and shows a maximum at Ni. A similar trend is observed for 4d metals. Figure 3 shows positron densities around the Ca13, Fe13 and Ni13 clusters. In the case of the Ca13 cluster, the annihilation fraction is 1.6 % and the positron density at the Ca13 cluster is close to zero. With increasing annihilation fraction from the Ca13 to Ni13 clusters, the positron density at the solute clusters becomes larger, which clearly indicates that the localization of positrons at the solute clusters increases with increasing annihilation fraction of the solute clusters.

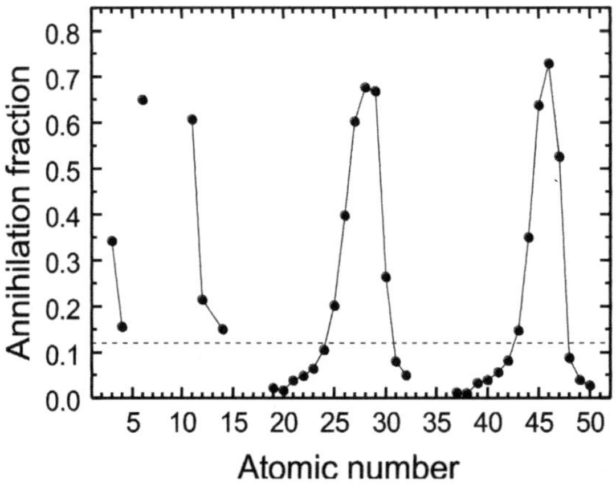

Figure 2. Annihilation fractions of the solute clusters in Al alloys as a function of the atomic number of the solute atoms.

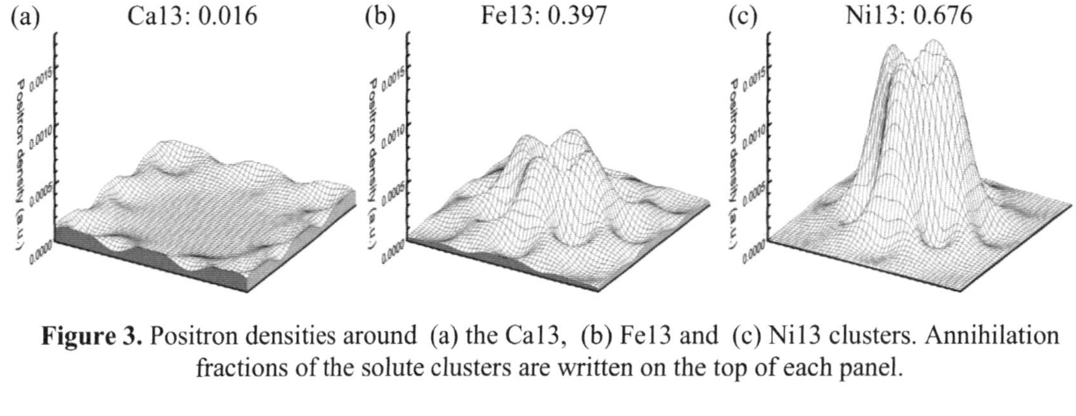

Figure 3. Positron densities around (a) the Ca13, (b) Fe13 and (c) Ni13 clusters. Annihilation fractions of the solute clusters are written on the top of each panel.

3.3. Potentials sensed by positrons and positron affinities

Although the annihilation fraction of the solute clusters directly indicates the decree of localization of positrons at the solute clusters, it does not have a physical meaning and strongly depends on the concentration or size of the solute clusters. Therefore, the annihilation fraction is not suitable for a parameter to estimate the positron affinities of the solute clusters.

The important factor to determine the distribution of positrons is the potentials sensed by positrons defined by equation (1). The potentials sensed by positrons for various sizes of the Zn and Ag clusters are shown in Figure 4. The lowest potentials sensed by positrons are not sensitive to the size of solute clusters. Thus, in the present work, we define the positron affinity of the solute clusters by the difference in the potentials sensed by positrons between the solute cluster and Al bulk as shown in Figure 5.

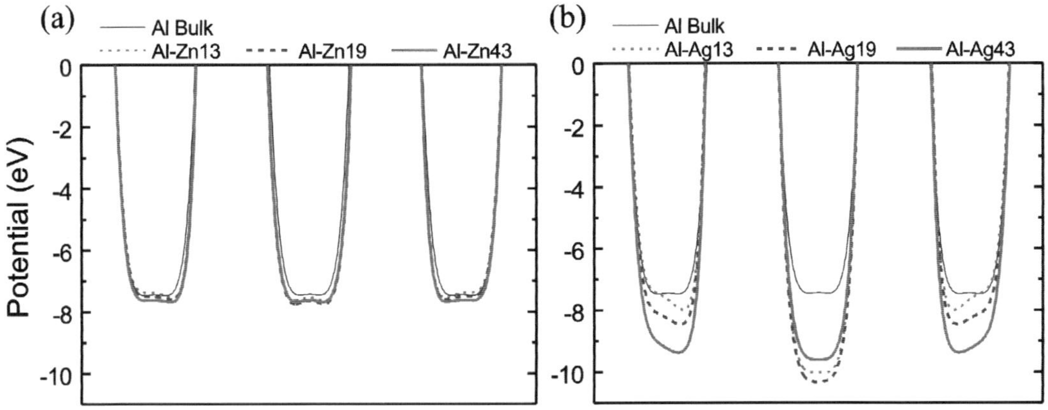

Figure 4. Potentials sensed by positrons around the solute clusters in (a) Al-Zn and (b) Al-Ag alloys.

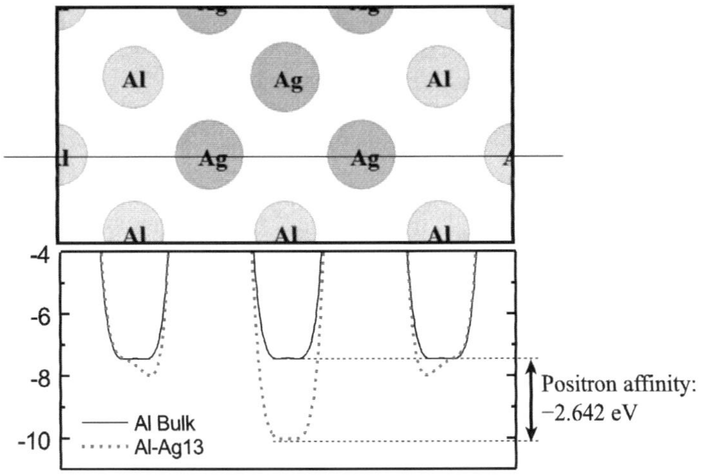

Figure 5. Definition of the positron affinity of the solute cluster.

3.4. Relationship between positron affinities and annihilation fractions

The positron affinities and annihilation fractions of the solute clusters are shown in Figure 6. In order to fit the scale to the annihilation fractions, the positron affinities are plotted from positive to negative values. The positron affinities show a similar trend to the annihilation fractions. Especially, the change in annihilation fractions in the 3d metals is well reproduced by the positron affinities. However, there is a significant difference on the left side of the 4d metals. For example, although the Mo cluster has almost the same positron affinity as the Co cluster, the annihilation fraction of the Mo cluster indicates that positrons are not localized at the Mo cluster. There is another difference in the magnitude between the positron affinities and annihilation fractions. The positron affinities of the 4d metals are larger than those of the 3d metals. However, the annihilation fractions are similar to each other. These discrepancies mainly arise from the difference in width of the potentials in the solute clusters.

Figure 7 shows the potentials sensed by positrons around the interstitial regions in the Co and Mo clusters and Al bulk. In the Co and Mo clusters, the depths of the potentials are almost the same values. However, the width of the potential in the Mo cluster is smaller than that of the Co cluster and Al bulk. As a result, the positrons are not localized at the Mo cluster. The difference in width of the potentials arises from the screening of the ion cores by d-electrons. The screening of the ion cores by 4d-electrons in the Mo clusters is weaker than that by 3d-electrons in the Co clusters. The depth and width of the potentials sensed by positrons in the 3d and 4d metal clusters are shown in Figure 8. The depth of the potential of the 4d metals is always larger than that of the 3d metals. On the other hand, the width of the potential of the 4d metals is always smaller than that of the 3d metals. Thus, the 3d and 4d metals show similar trends for the annihilation fractions despite the difference in the positron affinities. In order to improve accuracy of the positron affinities, not only the difference in depth of the potentials but also the difference in width of the potentials should be considered. We will try to define the positron affinity in which the width and depth of the potentials are taken into account in a future study.

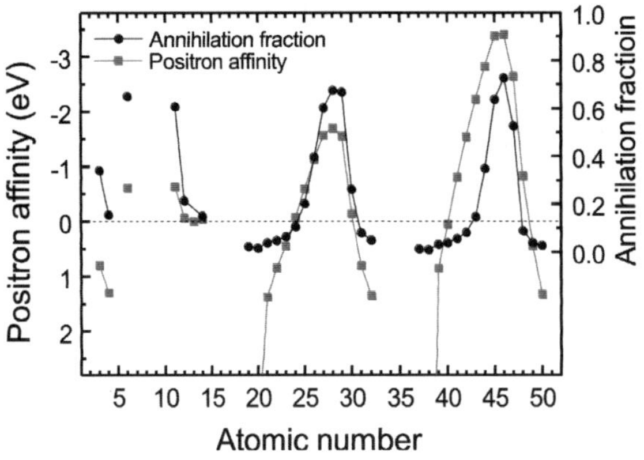

Figure 6. Correlation between positron affinities and annihilation fractions of solute clusters in Al alloys.

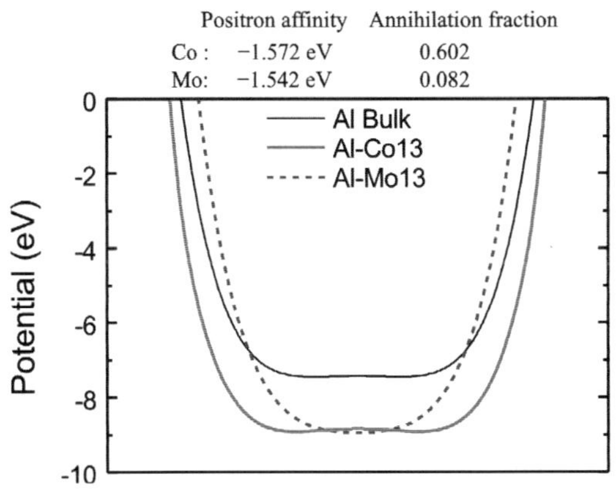

Figure 7. Potentials sensed by positrons around the Co13 and Mo13 clusters.

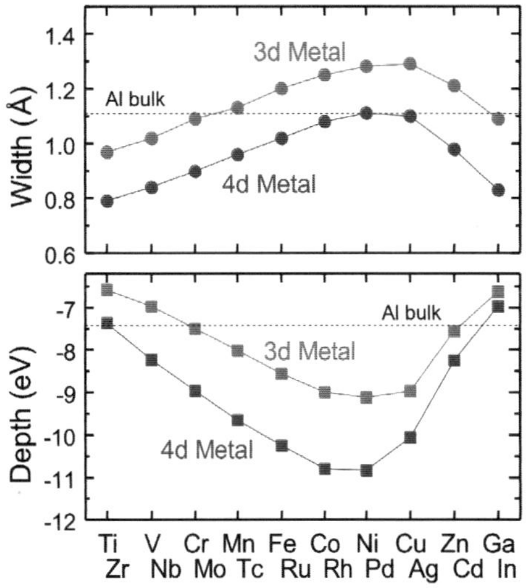

Figure 8. Widths and depths of potentials sensed by positrons around the solute clusters.

3.5. Difference in positron affinities between elemental metals and solute clusters in Al alloys

Figure 9 shows the positron affinities defined in the present work and those for elemental metals reported by Puska [3]. The positron affinities of elemental metals are shifted to set the value of Al to zero. The overall trend is quite different because the positron affinities for elemental metals are calculated for their own structures. A common feature is a decreasing trend in the middle of the 3d and 4d metals. The decreasing trend mainly arises from the increase in the screening of ion cores by d-electrons. The positron affinities for elemental metals show an increasing trend in the left side of the 3d and 4d metals. This trend is mainly due to the decrease in the atomic radius. The positron affinities for elemental metals have been used for estimation of positron affinities of solute clusters in Fe-Cu, [1] Al-Zn and Al-Ag [2]. In these alloys, the atomic radius of the solute atoms is comparable to that of the matrix atoms. Therefore, the positron affinities for elemental metals are applicable to the solute clusters in these alloys.

In order to clarify the origin of the positron affinities of the solute clusters, the positron affinities are compared to the charge transfers from Al bulk to the solute clusters in Figure 10. The change in the charge transfer is almost the same as that in the positron affinities. In the present work, the positron affinities are defined by the difference in the potentials sensed by positrons between the solute clusters and Al matrix. Thus, the difference in the potentials sensed by positrons mainly arises from the charge transfers from Al bulk to the solute clusters.

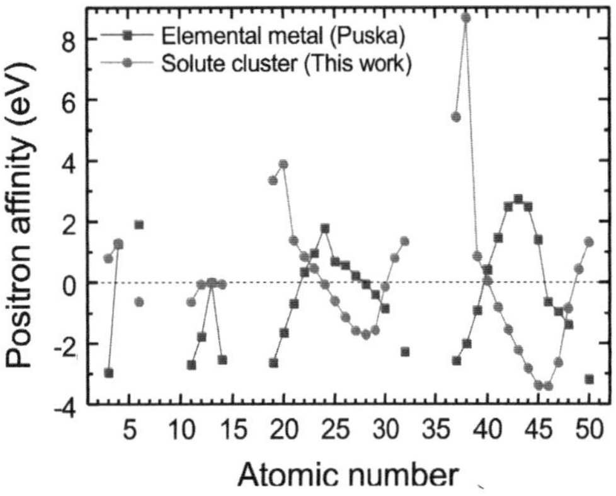

Figure 9. Comparison of positron affinities between elemental metals and solute clusters in Al alloys.

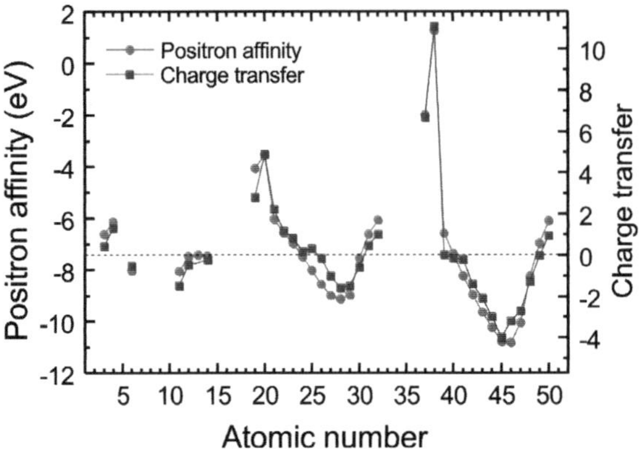

Figure 10. Correlation between positron affinities and net charges of solute clusters in Al alloys.

4. Conclusions

Positron affinities of solute clusters in aluminum have been calculated for Al-*X*13 clusters using first-principles molecular orbital calculations. Positron affinities defined by the difference in the potentials sensed by positrons well represent the localization of positrons at the solute clusters. Compared to the positron affinities of elemental metals, the overall trend in those of the solute clusters is quite different because the positron affinities for elemental metals are calculated for their own structures. A common feature is a decreasing trend in the middle of the 3d and 4d metals, which arises from the increase in the screening of ion cores by d-electrons. The difference in the potentials sensed by positrons is mainly due to the charge transfer from Al bulk to the solute clusters. The change in width of the potentials needs to be considered to improve accuracy of the positron affinities.

Acknowledgements

This study is partly supported by "Priority Assistance of the Formation of Worldwide Renowned Centers of Research–The Global COE Program (Project: Center of Excellence for Advanced Structural and Functional Materials Design)" and partly by a Grant-in-Aid for Scientific Research both from the Ministry of Education, Culture, Sports, Science and Technology of Japan.

References

[1] Nagai Y, Hasegawa M, Tang Z, Hempel A, Yubuta K, Shimamura T, Kawazoe Y, Kawai A and Kano F 2000 *Phys. Rev.* B **61** 6574
[2] Nagaia Y, Tang Z, Ohkubo H, Takadate K and Hasegawa M 2003 *Rad. Phys. Chem.* **68** 381
[3] Puska M J, Lanki P and Nieminen R M 1989 *J. Phys.: Condens. Matter 1* 6081
[4] Bharathi A and Chakraborty B 1988 *J. Phys. F: Met. Phys.* **18** 363
[5] Adachi H, Tsukada M and Satoko C 1978 *J. Phys. Soc. Jpn.* **45** 875
[6] Ellis D E and Painter G S 1970 *Phys. Rev.* B **2** 2887
[7] Mulliken R S 1955 *J. Chem. Phys.* **23** 1833
[8] Boroński E, Nieminen R M 1986 *Phys. Rev.* B **34** 3820
[9] Puska M J, Nieminen R M 1983 *J. Phys.* F **13** 333
[10] Mäder K A, Zunger A 1995 *Phys. Rev.* B **51** 10462

Investigation and calculation of positron lifetimes of mono-vacancies in crystals

Shijuan Huang[1,2], Jiandang Liu[1,2] and Bangjiao Ye[1,2]

[1] State Key Laboratory of Particle Detection and Electronics (IHEP&USTC), University of Science and Technology of China, Hefei 230026, China
[2] Department of Modern Physics, University of Science and Technology of China, Hefei 230026, China

E-mail: hsj2010@mail.ustc.edu.cn

Abstract. The first-principles calculations of positron lifetimes of mono-vacancies in crystals were investigated. We use the two-component density functional theory to respectively compute positron lifetimes of neutral charge state of V_{Al} defect in aluminium, V_{Si} defect in silicon, V_C, V_{Si} and V_C+C_{Si} defects in 3C silicon carbide, V_{Ga} and V_{As} defects in gallium arsenide, taking into account atomic relaxation due to vacancy and electronic structural relaxation due to the presence of the positron. Three different calculation schemes are used. We find that the electron density inside the vacancy more or less increases due to the presence of the positron if the ionic positions are kept fixed, and the positron becomes more localized after the electronic structural relaxation for the case of V_{Al} defect in aluminium and V_{Si} defect in 3C silicon carbide, but it is opposite for the case of V_{Ga} defect in gallium arsenide and V_C defect in 3C silicon carbide. The results with no consideration of the relaxation are even much closer to the experimental ones, therefore the atomic relaxation due to the position play an important role in calculating the positron lifetime of mono-vacancies in crystals.

1. Introduction

As is well known, positron annihilation spectroscopy (PAS) has been a powerful nondestructive technique to probe open volume defect in materials. As a positron gets trapped in a vacancy-type defect, the positron lifetime increases and this can be measured by the positron lifetime spectroscopy. Moreover the vacancy-type defect often has a very close relationship to materials' property, so it is required to identify the type of defects corresponding to a specific lifetime comparison with other characterization methods or calculated lifetimes. The interpretation of the positron lifetime spectroscopy experimental results is usually based on theoretical lifetime calculations, especially the bulk lifetime calculations.

Positron lifetime in perfect crystals has been demonstrated successfully by first principles calculation in past decades. They all almost obtain the similar results in the conventional way and the self-consistent way. While for the situation of positron in mono-vacancies of materials, there still exit some disputes. It has been reached that positron can be easily trapped and entirely located by negative or neutral vacancies, but different calculating results of positron lifetimes of vacancies are obtained by different authors. The discrepancies among those controversial results are mainly caused by neglecting the change of electron density and potential, which is usually related to atomic relaxation induced by vacancies and the trapped positron. Moreover, different methods applied in the calculation process are also lead to such discrepancies for positron lifetimes, such as superposed neutral atom model, pseudo-potential model, full-potential model, the local density approximation (LDA), the generalized gradient approximation (GGA), etc. Generally, in the conventional calculation scheme, the influence of the change of electron density and potential due to the presence of vacancies and positron was not taken into account, while these results agree with experimental ones much better than those results that all are considered in some situations, and this is the biggest dispute.

For the theoretical model, the electron pileup in the vacancy region due to the atomic relaxation induced by vacancies. Meanwhile, the presence of positron further attracts the electron around the vacancy region and repel atomic nucleus away from their balance position in the perfect crystal lattice, moreover, in the vacancy the positron density is not zero in fact and this diminishes contact rate, thus cancellation exists. Therefore further calculations and analysis need to be done carefully. Only a few of articles concerned about the fully-self-consistent scheme [1-4, 6]. Nieminen and Boronski etal [1,2] first proposed the two-component functional theory and applied it to calculate the positron lifetime of V_{Al} defect in aluminum, they found that the positron potential inside the vacancy became lower in self-consistent way than that in conventional way, and the positron distribution became more spread out. Puska etal [3] presented a self-consistent way to calculate the case of the ideal triply negative Ga vacancy in GaAs, the electron density inside the vacancy increased, and the positron became more localized as the trapped potential became lower. Julia Wiktor etal [4] used the pseudo-potential method to deal with the valence electrons, while the result they got was a little big different from the others, especially for the positron lifetime of V_C defect in SiC.

We present first-principle calculations of positron lifetimes in the vacancies in two schemes, one is the conventional way based on the one component density functional theory, and another is the self-consistent way based on the two component density functional theory. In the self-consistent way, we consider the influence induced by atomic relaxation due to the vacancy and the electronic structural relaxation due to the presence of the positron. In this paper, we choose Al, Si, GaAs and 3C-SiC as examples to calculate their corresponding mono-vacancy positron lifetimes, compare our results with the experiment lifetimes from the literatures.

2. Method

Theoretically, the positron annihilation rate λ takes the form related to the positron density $n_+(\vec{r})$, the electron density $n_-(\vec{r})$ and the enhancement factor $g_{12}(0;n_+(\vec{r});n_-(\vec{r}))$, and the positron lifetime τ is just the inverse of the positron annihilation rate:

$$\frac{1}{\tau} = \lambda = \pi r_0^2 c \int d\vec{r} n_+(\vec{r}) n_-(\vec{r}) g_{12}(0;n_+(\vec{r});n_-(\vec{r})), \qquad (1)$$

where c is the speed of the light and r_0 is the classical radius of an electron.

The electron and positron density can be calculated using various approximations, which leads to several schemes. In this study, we considered two schemes. One is the self-consistent way (SC) based on the two-component density-functional theory:

$$\begin{cases} -\frac{1}{2}\nabla^2\psi_i(\vec{r}) + [\frac{\delta E_{xc}[n_-]}{\delta n_-(\vec{r})} - \phi(\vec{r}) + \frac{\delta E_{corr}[n_+,n_-]}{\delta n_-(\vec{r})}]\psi_i(\vec{r}) = \varepsilon_i\psi_i(\vec{r}); \\ -\frac{1}{2}\nabla^2\psi_+(\vec{r}) + [\frac{\delta E_{xc}[n_+]}{\delta n_+(\vec{r})} + \phi(\vec{r}) + \frac{\delta E_{corr}[n_+,n_-]}{\delta n_+(\vec{r})}]\psi_+(\vec{r}) = \varepsilon_+\psi_+(\vec{r}). \end{cases} \qquad (2)$$

$$n_-(\vec{r}) = \sum_i |\psi_i(\vec{r})|^2, n_+(\vec{r}) = |\psi_i(\vec{r})|^2, \quad \phi(\vec{r}) = \int d\vec{r}' \frac{-n_-(\vec{r}') + n_+(\vec{r}') + n_0(\vec{r}')}{|\vec{r} - \vec{r}'|}. \qquad (3)$$

Where $\psi_i(\vec{r})$ and ε_i , $\psi_+(\vec{r})$ and ε_+ are respectively the electron wave function and its energy eigenvalue, the positron wave function and its energy eigenvalue, $\frac{\delta E_{xc}[n]}{\delta n(\vec{r})}$ is the electron or positron exchange-correlation potential, $\frac{\delta E_{corr}[n_+, n_-]}{\delta n(\vec{r})}$ is the electron or positron correlation potential, and $n_0(\vec{r})$ denotes the density of positive charge providing the external potential. In this scheme, the presence of the positron influence the distribution of the electrons and ions, conversely the change of the electronic structure influence the distribution of the positron, the ground electron density and the positron density are determined through the self-consistent solution of equations (2) and (3). In our situation, there is only one positron in the solid at a time, if we assume that the electron density isn't affected by the presence of the positron at all, we can obtain the density from the one-component density-functional functions:

$$\begin{cases} -\frac{1}{2}\nabla^2\psi_i(\vec{r}) + [-\phi(\vec{r}) + \frac{\delta E_{xc}[n_-]}{\delta n_-(\vec{r})}]\psi_i(\vec{r}) = \varepsilon_i\psi_i(\vec{r}); \\ -\frac{1}{2}\nabla^2\psi_+(\vec{r}) + [\phi(\vec{r}) + \frac{\delta E_{corr}[n_+, n_-]}{\delta n_+(\vec{r})}]\psi_+(\vec{r}) = \varepsilon_+\psi_+(\vec{r}). \end{cases} \quad (4)$$

$$\phi(\vec{r}) = \int d\vec{r}' \frac{-n_-(\vec{r}') + n_0(\vec{r}')}{|\vec{r} - \vec{r}'|} \quad (5)$$

The electron density first, then the positron density, this is the so-called conventional way (CONV).

And for the enhancement factor which taking into account an increase in the electron density at the positron site caused by the screening of the positron by electrons, it is a crucial ingredient in the context of calculating the positron lifetimes in solids, there are many different parameterized forms so far. In this study, we chose three forms, one is proposed by Boronski and Nieminen (BNLDA) [2], the second one is proposed by Barbiellini and others (APGGA) [5], and the last one is parameterized by Puska, Seitsonen and Nieminen (PSN) based on the result of Boronski and Nieminen [6].

3. Computational details and testing

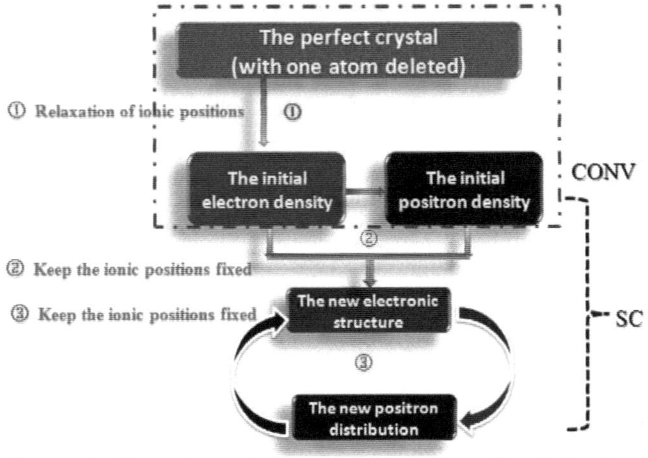

Figure 1. The schematic diagram of computational details.

We present a scheme as shown in figure 1 to calculate the positron lifetime for perfect crystals and crystals with vacancy-type defect. In the first step, we take a relaxation of ionic positions only due to the vacancy, then get the electron density as its initial values, then achieve the positron density by solving the positron part of one-component density-functional equations as its initial values. In the second step, we keep the ionic positions fixed and relax the electronic structure due to the presence of the positron, then solve the two-component density-functional equation to get the new positron distribution. At last, repeat the second step until convergence is reached. In the CONV scheme, only the first step is needed.

Positron lifetime calculation is performed using the Vienna Ab-initio Simulation Package (VASP) [7, 8]. For the relaxation in the first step, we use a conjugate-gradient algorithm. The force acting on each ion is less than 0.001eV/Angst. For the electronic structure, we adopt the PAW-PBE pseudo-potential [9] and plane wave basis set, and only take into account the valence electrons. The electron-electron exchange-correlation potential is performed using PBE-GGA method. And for the positron density, we use a real space grid method, the energy convergence is $1.0e^{-7}a.u.$.

For testing our program, we take the perfect crystals of Al, 3C-SiC, GaAs and Si as examples to calculate their bulk positron lifetimes. We use the experimental lattice parameters of 4.05 Å for Al, 4.33 Å for 3C-SiC and 5.43 Å for Si and theoretical lattice parameter of 5.57 Å for GaAs [3]. We take $2s^2 2p^2$, $3s^2 3p^1$, $3s^2 3p^2$, $4s^2 4p^1$ and $4s^2 4p^3$ as valence electrons for C, Al, Si, Ga and As respectively. The results are presented in Table 1.

Table 1. Comparison of the bulk positron lifetime (in ps) calculated in CONV and SC way. Different forms of the enhancement factor (BNLDA, APGGA and PSN) were used. The corresponding experimental values are obtained from the literatures.

Crystal	BNLDA		APGGA		PSN		Exp.
	CONV	SC	CONV	SC	CONV	SC	
Al	186	184	164	162	197	194	161[a]
3C-SiC	147	145	146	141	146	144	140[b]
GaAs	242	230	246	234	253	249	231[c]
Si	230	227	229	217	239	234	219[a]

[a] Corresponding experimental results within reference 10.
[b] Corresponding experimental results within reference 11.
[c] Corresponding experimental results within reference 6.

It can be noticed in Table 1 that the bulk positron lifetimes calculated in CONV way are similar to that calculated in SC way for these four perfect crystals, which is consistent with the previous studies. Of course, as we only take into account of the valence electrons, neglect the annihilation rate of the core electrons, the bulk lifetimes are almost larger than the experimental ones especially for the case of aluminium which is metal. The results especially in APGGA/SC scheme are close to the experimental values. This means that our program is reliable. In PSN scheme, since the enhancement factor is parameterized based on another different positron-electron correlation potential [6], in this work we just deal with the positron-electron correlation potential in APGGA scheme, the results are a little different from the experimental ones, and this shows that the positron-electron correlation potential plays an important role in calculating the positron lifetimes.

4. Results and discussion

For mono-vacancies in crystals, we select three schemes, NYH stands for no relaxation due to the vacancy and the presence of the positron, YH stands for only relaxation of the ionic positions due to the vacancy and YH+P stands for the relaxation of the ionic positions firstly, then the relaxation of the electronic structure due to the presence of the positron. Otherwise, all the mono-vacancies are handled as neutral ones.

Firstly, we study the relaxation of the ionic positions due to the vacancy. For all these crystals with vacancy-type defect, we use a cubic supercell with $2\times2\times2$ unit cells. In theoretically, the inward relaxation due to the vacancy should be obtained. However, after a careful test of convergence, there exits a very little outward expansion for the size of the supercell except in the case of the mono-vacancy in Al (see the Table 2). One reason may be that we use the experimental lattice parameters as initial values and they are always a little smaller than the theoretical ones. Another reason may be that the supercell size is not large enough. While the nearest neighbour distance becomes short after relaxation in YH scheme and this is reasonable. At the same time, we calculate the positron lifetimes of vacancy-type defect in Al, 3C-SiC, GaAs and Si in NYH and YH scheme. Next, we study the effect of the localized positron on the relaxation of the electronic structure surrounding a vacancy by calculating the electronic structure of monovacancies in bulk Al, 3C-SiC, GaAs and Si with and without the localized positron, meanwhile, the ionic positions are kept fixed.

Table 2. The calculated lattice constants (in Å) and relaxations for the crystals with vacancy-type defect in NYH and YH scheme. A positive (negative) number denotes the outward (inward) relaxation in percentage of the nearest neighbour distance with respect to the ideal vacancy.

	Al	$3C\_SiC$			$GaAs$		Si
	V_{Al}	V_{Si}	V_C	V_C+C_{Si}	V_{Ga}	V_{As}	V_{Si}
NYH	8.10	8.66	8.66	8.66	11.14	11.14	10.86
YH	8.06	8.75	8.74	8.81	11.46	11.48	10.90
Rel. (%)	-1.43	11.24	2.87	15.29	-8.85	-7.07	-6.97

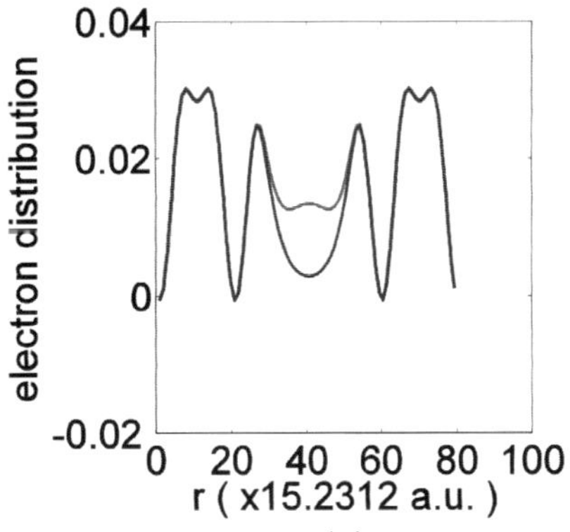

(a)

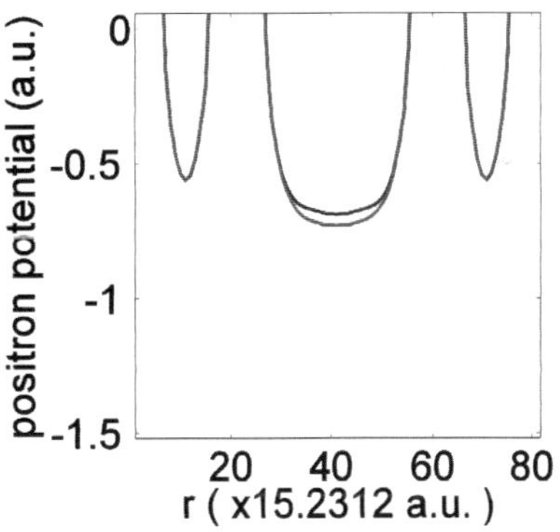

(b)

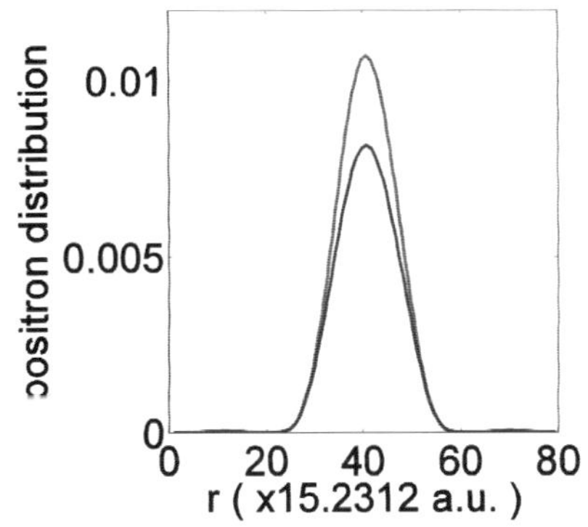

(c)

Figure 2. (color online) physical distribution of Al with monovacancy in [110] direction through the vacancy, the blue line denotes the distribution calculated in YH scheme and the red line denotes that calculated in YH+P scheme. (a) The electron density, (b) the trapping potential, (c) the positron density.

Journal of Physics: Conference Series **674** (2016) 012022 doi:10.1088/1742-6596/674/1/012022

For mono-vacancy in aluminum, shown in Figure 2, the electron density inside the vacancy slightly increases, the positron potential becomes deeper, and the positron is more localized in YH+P scheme than that in YH scheme, so the positron lifetime becomes shorter further.

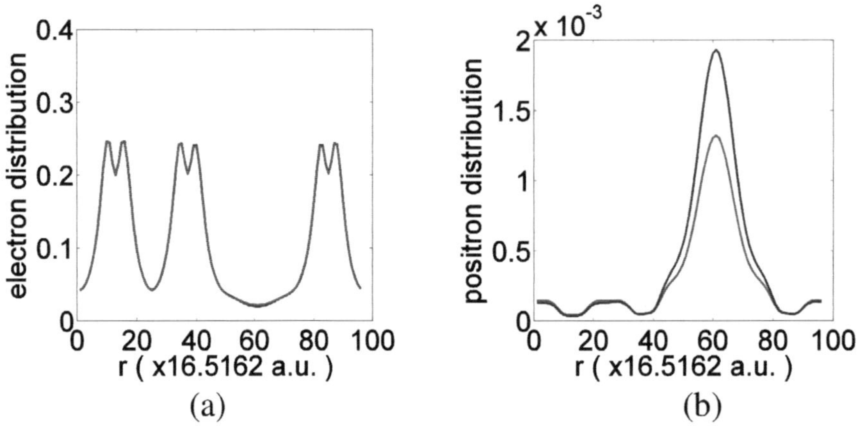

Figure 3. (color online) physical distribution of 3C-SiC with monovacancy of carbon in [110] direction through the vacancy, the blue line denotes the distribution calculated in YH scheme and the red line denotes that calculated in YH+P scheme. (a) The electron density, (b) The positron density.

And for carbon mono-vacancy in SiC, its positron lifetime is controversial. In some previous theoretical studies, it was found that the carbon defect is not a positron trapping state. In our study, the electron density inside the vacancy almost remains unchanged and the positron becomes less localized, as shown in Figure 3 (a) and (b) respectively. The calculated positron lifetime is very close to its bulk lifetime, so we think the carbon vacancy can not be detected by the positron lifetime spectroscopy.

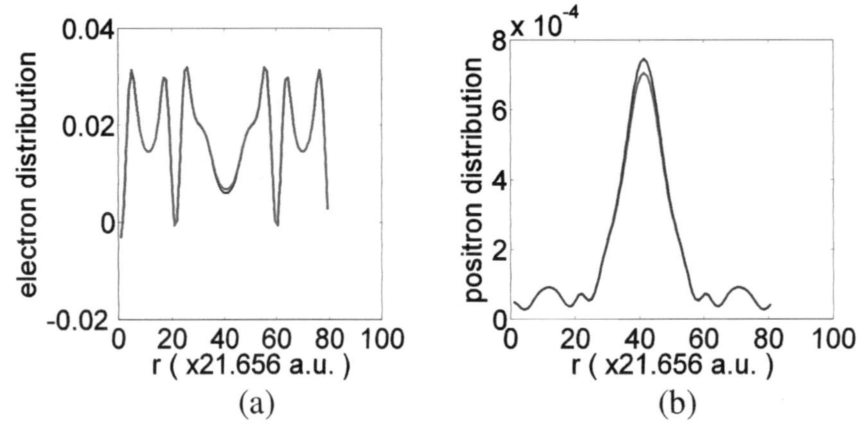

Figure 4. (color online) physical distribution of GaAs with monovacancy of gallium in [110] direction through the vacancy, the blue line denotes the distribution calculated in YH scheme and the red line denotes that calculated in YH+P scheme. (a) The electron density, (b) the positron density.

We also show the change of the electron and positron density inside the vacancy for the gallium vacancy in GaAs. As shown in Figure 4 (a) and (b), the electron density is just a little larger in YH+P scheme than that in YH scheme, and the positron is a little less localized. But in NYH scheme, the positron is well localized, as shown in Figure 5.

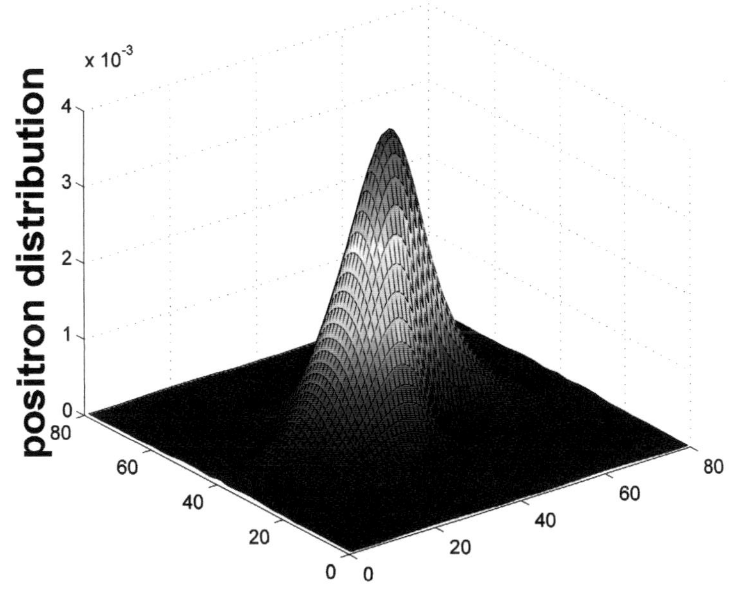

Figure 5. (color online) the distribution of the positron density of GaAs with monovacancy of gallium in NYH scheme in [110] direction.

Table 3. Comparison of the monovacancy positron lifetime (in ps) calculated in NYH, YH and YH+P way. Different forms of the enhancement factor (BNLDA, APGGA and PSN) were used. The corresponding experimental values are obtained from the literatures.

Crystal	NYH	YH	YH+P	Exp.
	BNLDA/APGGA/PSN	BNLDA/APGGA/PSN	BNLDA/APGGA/PSN	
Al_V_{Al}	242/219/276	235/212/265	221/194/243	244[a]
$3CSiC\_V_{Si}$	187/191/192	214/219/225	198/188/203	188[b]
$3CSiC\_V_C$	147/146/142	152/152/146	150/149/145	No[c]
$V_C + C_{Si}$	144/143/138	150/150/145	150/149/144	---
$GaAs\_V_{Ga}$	263/274/276	253/259/246	252/257/245	260[d]
$GaAs\_V_{As}$	257/268/268	254/261/248	254/259/247	257[e]
Si_V_{Si}	254/252/259	235/233/230	233/230/227	282[f]

[a] Corresponding experimental results within reference 12.
[b] Corresponding experimental results within reference 11.
[c] Corresponding experimental results within reference 13.
[d] Corresponding experimental results within reference 14.
[e] Corresponding experimental results within reference 15.
[f] Corresponding experimental results within reference 16.

We give the calculated positron lifetimes in different schemes and their corresponding experimental values in Table 3. The positron lifetimes do not change very much after we take the relaxation of the ionic positions due to the vacancy and the electronic structure due to the presence of the positron. The positron lifetimes in NYH scheme are even much closer to the experimental ones, especially for the vacancy in silicon, we conclude that the effect of the relaxation of the electronic structure due to the positron may be less important than that of the relaxation of the ionic positions due to the positron in some cases. The studies on silicon with monovacancy [17-19] in fully self-consistent way give lifetimes around 270ps, this is very close to the experimental one, and clearly prove that the effect of the relaxation due to the vacancy can not be compensated by that of the relaxation due to the presence of the positron.

5. Conclusions

The positron lifetimes of monovacancies in crystals are studied in the conventional way and self-consistent way. For perfect crystals, the similar results are obtained in these two ways, and the results in the GGA form are much closer to the experimental ones. For monovacancies in crystals, we select three schemes to calculate the positron lifetimes. In YH+P scheme, the electron density inside the vacancy more or less increases due to the presence of the positron if the ionic positions are kept fixed, and after the relaxation of the electronic structure, the positron is more localized for the mono-vacancy in aluminum, while less localized for the Gallium mono-vacancy in GaAs and Carbon mono-vacancy in SiC. The results with no consideration of the relaxation are even more similar with the experimental ones, especially for silicon monovacancy in silicon, so we conclude that maybe the effect of the relaxation of the electronic structure due to the positron is less important than that of the relaxation of the ionic positions due to the presence of the positron in some cases, in other words, the effect of the relaxation due to the vacancy can not be compensated by that of the relaxation due to the presence of the positron in some cases.

Acknowledgement

This study was partially financially supported by National Natural Science Foundation of China (No. 11175171, 11105139). The authors are grateful to Wenshuai Zhang for insightful discussion.

References

[1] Nieminen R M , Boronski E and Lantto L J 1985 *Phys. Rev.* B **32** 1377
[2] Boronski E and Nieminen R M 1986 *Phys. Rev.* B **34** 3820
[3] Gilgien L, Galli G, Gygi F and Car R 1994 *Phys. Rev.* L **72** 3214
[4] Wiktor J, Jomard G, Torrent M and Bertolus M 2013 *Phys. Rev.* B **87** 235207
[5] Barbiellini B, Puska M J, Torsti T and Nieminen R M 1995 *Phys. Rev.* B **51** 7341
[6] Puska M J, Seitsonen A P and Nieminen R M 1995 *Phys. Rev.* B **52** 10947
[7] Kresse G and Hafner J 1993 *Phys. Rev.* B **47** R558
[8] Kresse G and Furthmuller J 1996 *Phys. Rev.* B **54** 169
[9] Blochl P E 1994 *Phys. Rev.* B **50** 17953
[10] Campillo-Robles J M, Ogando E and Plazaola F 2012 *Solid State Science* **14** 982
[11] Kawasuso A, Itoh H, Morishita N, Yoshikawa M, Ohshima T, Nashiyama I, Okada S, Okumura H and S. Yoshida 1998 *Appl. Phys.A* **67** 209
[12] Robles J M C and Plazaola F 2003 *Defects and Diffusion in Metals* **213-2** 141
[13] Kawasuso A, Yoshikawa M, Itoh H, Krause-Rehberg R, Redmann F, Higuchi T and Betsuyaku K 2006 *Physica.* B **376-377** 350
[14] Corbel C, Pierre F, Saarinen K, Hautojarvi P and Moser P 1992 *Phys. Rev.* B **45** 3386
[15] Saarinen K, Hautojarvi P, Lanki P and Corbel C 1991 *Phys. Rev.* B **44** 10585
[16] Polity A, Borner F, Huth S, Eichler S and Krause-Rehberg R 1998 *Phys. Rev.* B **58** 10363
[17] Saito M and Oshiyama A 1996 *Phys. Rev.* B **53** 7810
[18] Makhov D V and Lewis L J 2005 *Phys. Rev.* B **71** 205215
[19] Makkonen I, Hakala M and Puska M J 2006 *Phys. Rev.* B **73** 035103

Monte Carlo analysis of germanium detector performance in slow positron beam experiments

J. Heikinheimo, R. Tuominen and F. Tuomisto

Aalto University, P.O. Box 14100, FI-00076 AALTO, Finland

E-mail: janne.heikinheimo@aalto.fi

Abstract. Positron annihilation Doppler broadening spectroscopy is one of the most popular positron annihilation vacancy characterization techniques in experimental materials research. The measurements are often carried out with a slow positron beam setup, which enables depth profiling of the samples. The key measurement devices of Doppler broadening spectroscopy setups are high-purity germanium detectors. Since Doppler broadening spectroscopy is one of the standard techniques in defect characterization, there is a demand to evaluate different kinds of factors that might have an effect on the results. Here we report the results of Monte Carlo simulations of detector response in different geometries and compare the data to experiments.

1. Introduction

The Monte Carlo N-particle transport code (MCNP) is one of the most used programs to simulate germanium detector properties [1, 2]. Monte Carlo simulations are a good way to identify and model small details in the detector response that are otherwise difficult or impossible to detect in experimental setups. Monte Carlo simulations have been used earlier, too, to model interactions of annihilation radiation in Doppler broadening spectroscopy. This was done for two HPGe detectors in coincidence, and the goal was to recognize pile-up effects in the measurement setup [3].

The peak efficiency of the germanium detector for the annihilation line is one of the important detector parameters in Doppler broadening measurements. The optimal dimensions for the Ge crystal are different for 511 keV gamma photons than for example for 1332.5 keV photons emitted by Co-60, which is commonly used for determining the detector efficiency. The optimal dimensions for efficiencies of germanium detectors with a specific volume were obtained by simulating coaxial detectors with different diameters and thicknesses with typical annihilation-detector geometry. The effect of the thickness of a steel plate placed between the photon source and the detector was analyzed by comparing measurements with different steel thicknesses to the corresponding simulations with the 511 keV gamma line alone.

To simulate the final spectrum of the Doppler broadening radiation, ab initio theoretical predictions of the electron-positron momentum distribution were used as source distribution. Here, GaN bulk and Ga vacancies in GaN were used for analyzing how different factors in the measurement setup affect the final results [4, 5].

2. Methods

2.1. Model for MCNP simulations

Generally, measured and simulated results correspond well to each other for high purity germanium detectors. The difference in detector efficiency could still be of the order of 10 %. Monte Carlo methods tend to overestimate the detector efficiency in the model based on the information from the suppliers [6, 7]. The difference could be caused by thicker dead layer or transition zone in germanium crystal, which changes the dimensions of the active region. The correspondence between the simulated and experimental results has been found by carefully adjusting the dead layer and crystal dimensions [8, 9, 10].

In this work, the dimensions for the germanium detector, crystal, and dead layers are obtained from the technical data provided by manufacturer (Canberra Industries, Inc.). The dimensions for the detector that has been used also in the Monte Carlo model are described in figure 1.

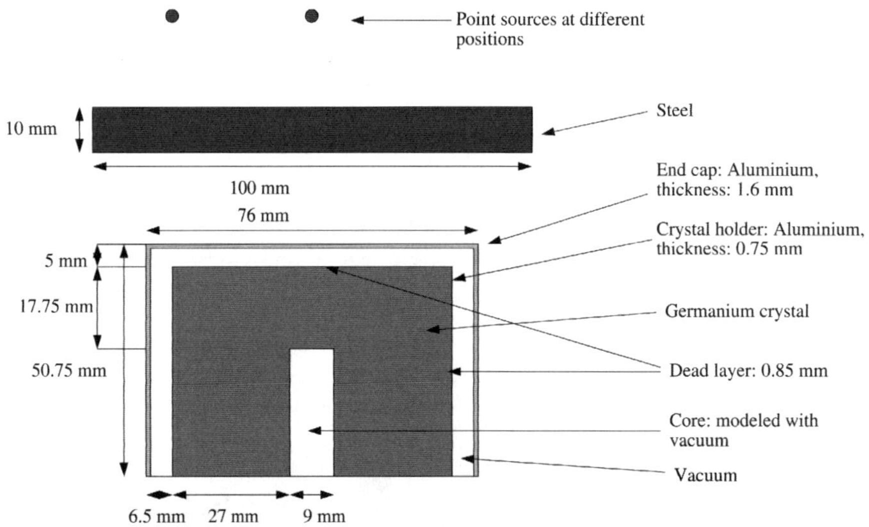

Figure 1. A scheme of the model used in the Monte Carlo simulations. Germanium detector dimensions have been chosen according to the detector that was used in the experiments.

2.2. Photon interactions in MCNP program

The physical treatment for photons includes the photoelectric effect, pair production, incoherent (Compton) scattering, coherent (Thompson) scattering, and fluorescent photons after photoelectric absorption. The electrons are treated in this work according to a thick-target bremsstrahlung model (TTB), where electrons travel in the direction of the incident photon and are immediately stopped.

In photoelectric effect, a photon gets absorbed completely giving its energy to release one electron in an atom and to the kinetic energy of the electron. Recombination effects in the exited atom are also taken into account. These effects could result as fluorescent photons and/or Auger electron.

In pair production, the energy of the photon goes to the creation of the electron-positron pair and to their kinetic energies. The kinetic energy of the electron-positron pair is

$$E_{\pm} = E_p - 2mc^2, \tag{1}$$

where E_p is the initial photon energy and mc^2 is the rest energy of an electron. In this work, positrons and electrons from the pair production were treated according to TTB. When positron energy is below electron energy cutoff (defined in MCNP), it is considered to be annihilated with an electron producing two photons going to the opposite directions with an energy of $mc^2 = 511$ keV.

In Compton scattering, a photon is scattered from a free electron. In this event, the objective is to determine the energy of the scattered photon E' and $\mu = \cos(\theta)$ for the angle θ of deflection from the line of flight. The differential cross section for the process is given by the Klein-Nishina formula

$$K(\alpha,\mu)\mathrm{d}\mu = \pi r_0^2 \left(\frac{\alpha'}{\alpha}\right)^2 \left[\frac{\alpha'}{\alpha} + \frac{\alpha}{\alpha'} + \mu^2 - 1\right] \mathrm{d}\mu, \tag{2}$$

where r_0 is the classical electron radius and α and α' are the incident and final photon energies in units $\alpha = E/mc^2$ and $\alpha' = \alpha/(1 + \alpha(1 - \mu))$. MCNP uses the modified Klein-Nishina formula for incoherent scattering

$$\sigma_I(Z,\alpha,\mu)\mathrm{d}\mu = I(Z,v)K(\alpha,\mu)\mathrm{d}\mu, \tag{3}$$

where $I(Z,v)$ is an appropriate scattering factor and $v = \sin(\theta/2)/\lambda$.

Thomson scattering involves no energy loss, and so is only a photon process which doesn't produce electrons. The differential cross section for Thomson scattering is

$$\sigma_2(Z,\alpha,\mu)\mathrm{d}\mu = C^2(Z,v)T(\mu)\mathrm{d}\mu, \tag{4}$$

where $C(Z,v)$ is a form factor and $T(\mu) = \pi r_0^2(1 + \mu^2)\mathrm{d}\mu$ is the energy-independent Thomson cross section.

3. Results

3.1. Optimization of the Ge crystal dimensions for detector efficiency

It is not always possible or even necessary to have custom designed detectors for slow positron beam measurements. If the detector resolution, peak-to-Compton ratio, and efficiency are sufficient for Doppler broadening measurements, the detector will suite well for its purpose even though its dimensions are not optimal for 511 keV gamma measurements. However, high-purity germanium is challenging and expensive to manufacture and the volume of the high-purity germanium crystal might show up in the detector price tag. Greater volume also generally increases detector capacitance and deteriorates energy resolution.

The optimization was performed with the source in the central position and 2 cm away from the end cap of the detector. The efficiency optimization was done for three different germanium volumes with a coaxial geometry. The results from the iteration are illustrated in figure 2. For $V = 80$ cm^3, the optimum was found when diameter was 6.8 cm and thickness 2.2 cm. Correspondingly, for $V = 142$ cm^3, the diameter was 8.2 cm and thickness 2.7 cm, and for $V = 200$ cm^3, the diameter was 9.2 cm and thickness 3.0 cm. For differently sized crystals, the optimal dimensions were found when the diameter-to-thickness ratio was around 3.

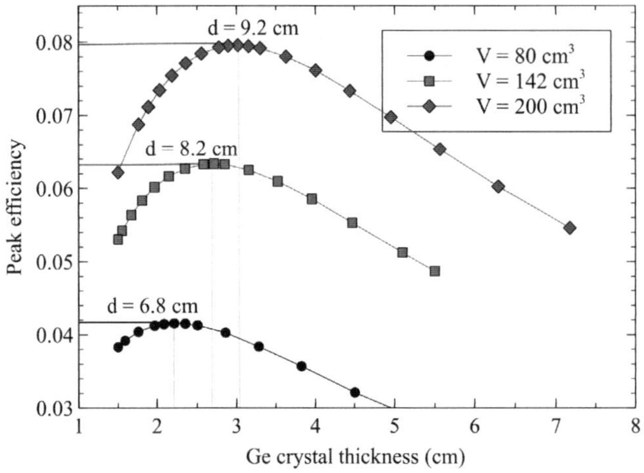

Figure 2. Detector efficiency optimization for three constant volumes.

3.2. Steel thickness and background effects

In positron laboratories, the most common vacuum material is stainless steel (with low magnetic permeability). Here, steel plates with different thickness were placed in between the source and the detector. The source was placed in the experiments and simulations at the central position and 5 cm away from the end cap of the detector

The source in this experiment was Na-22 with thin aluminum wrapping, and its activity was defined with the coincident method to be 78 ± 4 kBq. It was packed to the familiar sandwich package normally used in positron lifetime measurements. The sample material was p-type Si(100).

Figure 3 shows experimentally obtained results for different steel thicknesses. The spectra have been normalized to the same value, or more specifically, the S-regions have been normalized. The spectra behave expectedly in the sense that steel is known to be a rather good shielding material to gamma rays. So, when S-regions are normalized, the background level increases. Especially Compton scattering in steel can potentially increase background in the low-energy side right next to the annihilation peak. It should also be noted that Na-22 daughter nuclide (Ne-22) emits a 1.28 MeV gamma photon. This increases the background in the high-energy region and the effect of steel is visible also there. It is clear that the greater background caused by steel increases the error for *W*-parameter. The effect of steel on the *S*-parameter should be minor.

In the Monte Carlo simulations, the source-detector setup was similar otherwise but the positron annihilation was modeled as a 511 keV point source i.e. without Doppler broadening. The background radiation was not included to the source, either. The results for simulations with different steel thicknesses (modeled with pure iron) are shown in figure 4. When annihilation peaks are normalized to the same value, one can immediately see from the spectra how much the lower energy background next to the annihilation peak is affected by the Compton scattered photons in the steel.

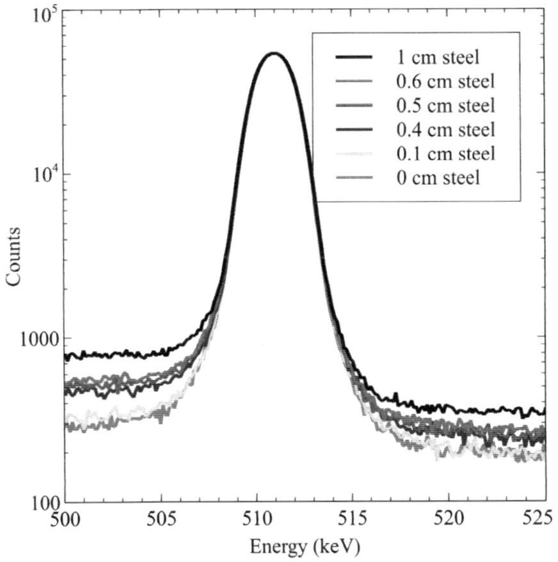

Figure 3. The effect of steel on the annihilation peaks, when the S-parameter region is normalized to the same value.

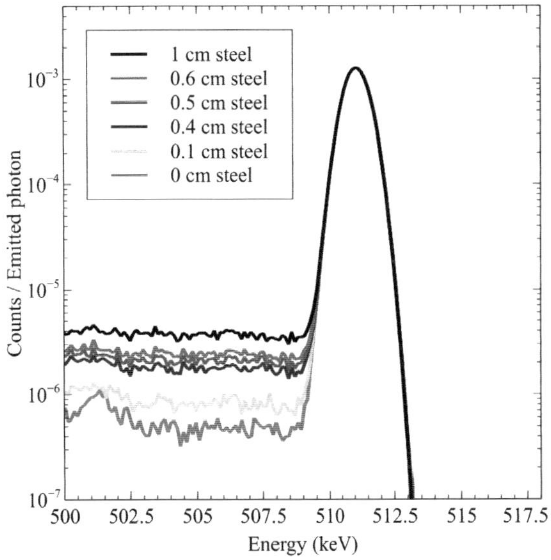

Figure 4. Results of Monte Carlo simulations for the germanium detector, when steel plates of different thicknesses were placed between the source and the detector. The steel plates were modeled in these simulations with pure iron.

3.3. Monte Carlo simulated cases for GaN and Ga vacancy in GaN

In order to get simulated results for Doppler broadening spectrum, results from first principles calculations were applied as a source distribution in the MCNP program. Figure 5 presents a measured ratio curve for the Ga vacancy in GaN [11]. It is followed by results obtained with the first principles calculations and finally the Monte Carlo simulated curve [4]. Here the detector resolution was 1.35 keV at 511 keV. Only the Compton background was taken into account in the background reduction since incomplete charge collection or pile up effects don't exist in the simulations.

As one can see from the figure 5, detector cover or other small details described in figure 1 (without steel) don't have an effect on the gamma interactions so that the ratio curve would change. Changing the source position to 3.15 cm off the axis and keeping the distance from the end cap didn't result in any changes in the ratio curve (the anisotropy of the electron-momentum distribution is not taken into account). Also placing 0.5 cm and 1 cm iron in between the source and the detector didn't result in changes big enough to be clearly separable from the original Monte Carlo simulated spectrum. Iron effect on the ratio curve is presented in figure 6. Changing germanium crystal dimensions didn't show any changes in the ratio curve, either. In all the cases, the *S*- and *W*-parameters remained unchanged.

It is known that the resolution changes the ratio curves. In the Monte Carlo simulations, resolution was applied by sampling the Gaussian distribution. Clear changes in the ratio curves and *S*- and *W*-parameters are obtained when the resolution changes 0.3 keV from the original value. The results are displayed in figure 7.

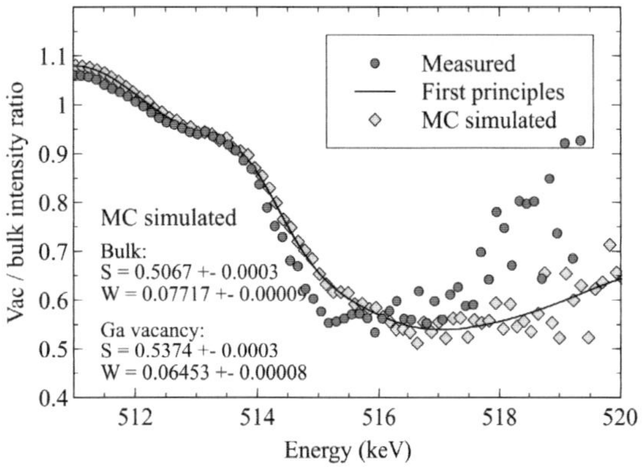

Figure 5. Ratio curves for Ga vacancy in GaN. In the Monte Carlo simulation, the source was placed at the central position 2 cm away from the end cap of the detector. Experimental data is from [11].

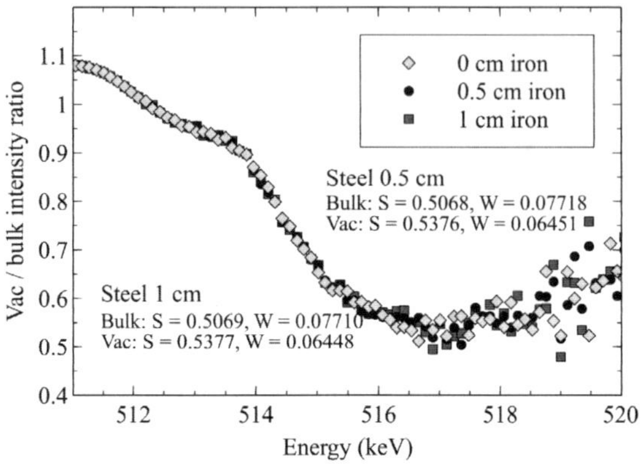

Figure 6. Iron effect on the Ga vacancy in GaN ratio curve. Iron plate is placed in between the source and the detector.

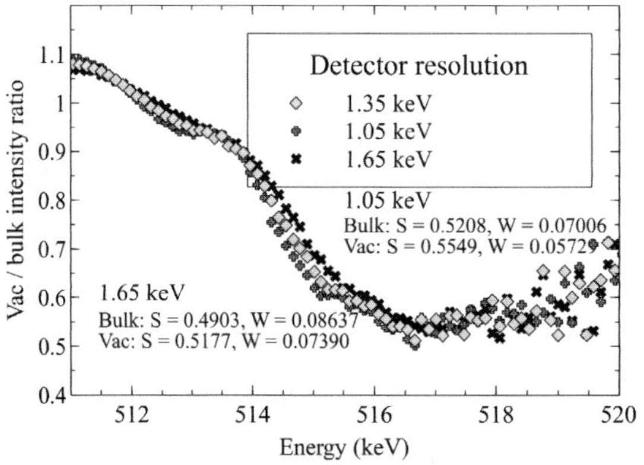

Figure 7. Ratio curves for Ga vacancy in GaN when resolution is varied.

4. Conclusions

With the Monte Carlo simulations, many detector properties could be simulated including final Doppler broadening spectra. However, one should bear in mind that the simulations show only energy deposited in the germanium crystal and not charged particle collection to the electrodes, for example. Here, the detector efficiency was optimized by changing the coaxial detector dimensions. With the three different volumes, the diameter-thickness ratio was close to 3 when the source is placed 2 cm away from the end cap of the detector.

Also background caused by excess steel in between the source and the detector was studied and compared to the corresponding experimental results. With the Monte Carlo simulations, one can conclude how much small angle Compton scattering in the steel increases background level in the low energy side of the annihilation peak.

Comparing the simulated ratio curves for Ga vacancy in GaN, one can study whether the measurement conditions change the final results significantly. In this study, only change in the detector resolution showed clear changes in the ratio curve and in the S- and W-parameters.

The Monte Carlo results obtained in this work could be applied to normal Doppler broadening measurements and 1-D coincidence measurements.

References

[1] J. F. Briesmeister (Editor) 2000 *MCNP^{TM} - A General Monte Carlo N-Particle Transport Code Version 4C Manual* (Los Alamos: National Laboratory)
[2] R. Berndt and P. Mortreau 2012 *Nucl. Instr. and Meth. in Phys.* A **694** 341
[3] E. do Nascimento, J. M. Fernandez-Varea, V. R. Vanin and N. L. Maidana 2011 *AIP Conference Proceedings* **1351** 216
[4] I. Makkonen and M. Puska 2007 *Phys. Rev.* B **76** 054119
[5] F. Tuomisto and I. Makkonen 2013 *Rev. Mod. Phys.* **85** 1583
[6] J. Rodenas, A. Pascual, I. Zarza, V. Serradell, J. Ortiz and L. Ballesteros 2003 *Nucl. Instr. and Meth. in Phys.* A **496** 390
[7] M.C. Lepy, T. Altzitzoglou, D. Arnold, F. Bronson, R.Capote Noy, M. Decombaz, F. De Corte, R. Edelmaier, E.Herrera Peraza, S. Klemola, M. Korun, M. Kralik, H. Neder, J. Plagnard, S. Pomme, J. de Sanoit, O. Sima, F. Ugletveit, L. Van Velzen and T. Vidmar 2001 *Appl. Radiat. Isot.* **55** 493
[8] J. Boson, G. Ågren, L. Johansson 2008 *Nucl. Instr. and Meth. in Phys.* A **587** 304
[9] C. Agarwal, S. Chaudhury, A. Goswami, M. Gathibandhe, 2011 *J. Radioanal. Nucl. Chem.* **287** 701
[10] N.Q. Huy, D.Q. Binh, V.X. An 2007 *Nucl. Instr. and Meth. in Phys.* A **573** 384
[11] S. Hautakangas, I. Makkonen, V. Ranki, M. J. Puska, K. Saarinen, X. Xu, and D. C. Look 2006 *Phys. Rev.* B **73** 193301

Defect studies of Mg films deposited on various substrates

P Hruška[1], J Čížek[1], W Anwand[2], J Bulíř[3], J Drahokoupil[3], J Stráská[1], O Melikhova[1], I Procházka[1], J Lančok[3]

[1] Faculty of Mathematics and Physics, Charles University in Prague, V Holešovičkách 2, 180 00 Prague 8, Czech Republic
[2] Institute of Radiation Physics, Helmholtz-Zentrum Dresden-Rossendorf, PO Box 510119, D 01314 Dresden, Germany
[3] Institute of Physics, Academy of Sciences of the Czech Republic, Na Slovance 2, 182 21 Prague 8, Czech Republic

E-mail: peta.hruska.1@gmail.com

Abstract. In the present work the structure of Mg films deposited by RF magnetron sputtering was characterized using variable energy positron annihilation spectroscopy combined with scanning electron microscopy and X-ray diffraction. The effect of deposition parameters, namely temperature, type of substrate and deposition rate, on the microstructure was examined. All Mg films studied grow with the basal (0001) plane parallel with the substrate and exhibit only negligible in-plane stress. Films deposited at room temperature are characterized by nanocrystalline structure with high volume fraction of grain boundaries. and positrons are preferentially trapped in open volume defects present at grain boundaries. In these films positrons are trapped predominantly in open-volume defects present at grain boundaries. With increasing deposition temperature the mean grain size increases and the volume fraction of grain boundaries decreases. Hence, in Mg films prepared at elevated temperatures positrons are trapped mainly at misfit dislocations compensating different atomic spacing in the films and the substrate. Moreover, it was found that slow deposition rate leads to higher density of defects compared to fast deposition rate. By annealing of Mg film with thin 20 nm Pd over-layer at 300°C for 1 hour Pd layer is mixed with Mg film forming a Mg-Pd compound. The Mg-Pd phase likely contains structural open-volume defects which trap positrons.

1. Introduction

Magnesium hydride (MgH_2) is an attractive material suitable for on-board hydrogen storage in mobile applications due to its safety, lightweight, low cost and in particular high hydrogen absorption capacity up to 7.6 wt. % [1]. However sluggish hydrogen sorption kinetics in MgH_2 is the major obstacle for its use in practical applications. Structure refinement down to nanoscale forms a dense network of grain boundaries which enable easier hydrogen distribution into grains and thus could remarkably speed up hydrogen sorption kinetics. The effect of microstructure on the hydrogen sorption kinetics can be investigated on thin films since samples with various microstructure from nanocrystalline to epitaxial can be relatively easily prepared by varying the parameters of the deposition. Structural studies of such Mg thin films are necessary as a first step of such investigation.

Variable energy positron annihilation spectroscopy (VEPAS) provides depth resolved information about defect structure in the studied material, hence it is a very useful tool for studies of the microstructure of thin Mg films [2-5].

2. Experimental

All Mg films were prepared in a vacuum chamber with pre-deposition pressure of ~10^{-4} Pa by RF (13.6 MHz) magnetron sputtering using an Ar atmosphere with a pressure of 3 Pa. Two different substrates were used: sapphire (0001) single crystal and amorphous fused silica (FS). Each film was immediately after deposition covered with 20 nm thick Pd over-layer in order to prevent oxidation of the Mg layer on the air and facilitating hydrogen absorption during electrochemical charging with hydrogen [6]. Four different deposition conditions were applied: (i) deposition at room temperature (RT), (ii) deposition on a substrate heated at 300°C, (iii) deposition at RT (including Pd cap) and subsequent annealing at 300°C for 1 h in the deposition chamber in Ar atmosphere with a pressure of 3 Pa and (iv) deposition at RT with slow deposition rate. Thicknesses of prepared films and deposition conditions are listed in Table 1.

Structural studies of Mg films were performed using FEI Quanta 200 scanning electron microscope (SEM). X-ray diffraction (XRD) measurements using Co K_α wavelength of 0.179026 nm were performed in Bragg-Brentano symmetric geometry on a PANalytical X'Pert Pro diffractometer.

VEPAS measurements were performed using a magnetically guided variable energy positron beam "SPONSOR" [7] with slow positron energies varying from 30 eV to 35 keV. Energy spectra of annihilation gamma rays were measured by HPGe detectors having a relative efficiency of ~30% and an energy resolution of 1.06(1) keV (FWHM at 511 keV). For each positron energy the Doppler broadening of annihilation profile was evaluated using the S parameter. The dependence of the S parameter on the positron energy was analyzed using the VEPFIT code [8].

3. Results and discussion

Figures 1 and 2 show SEM micrographs of the Mg films deposited on sapphire and FS substrate respectively. Mg films exhibit nanocrystalline structure, the mean grain sizes were estimated by SEM and are listed in Table 1. The films deposited on amorphous FS substrate exhibit larger grains compared to the films deposited on crystalline sapphire substrate using the same deposition conditions. This is most probably connected with lattice misfit between the substrate and the Mg film which is more pronounced for the single crystalline sapphire substrate than for the amorphous FS substrate.

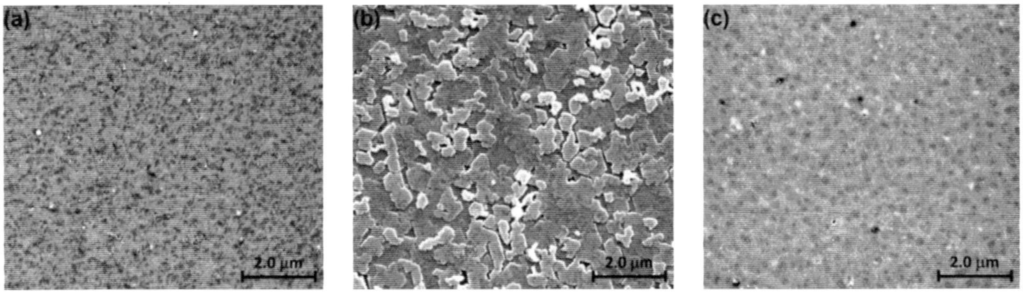

Figure 1. SEM micrographs of Mg films deposited on sapphire (0001) substrate at RT (a), at 300°C (b) and at RT and subsequently annealed at 300°C for 1 h (c).

According to the XRD measurements all Mg films are strongly textured independently on the type of substrate and the deposition condition. Mg films studied grew preferentially with the most densely populated basal plane (0001) parallel with the substrate, see Figure 3. The out-of-plane lattice parameter c measured in the Mg films differs from the tabular value of 0.52108 nm by less than 0.1%. This testifies to negligible in-plane stress in all Mg films studied. The lattice misfit between the film and the substrate is compensated by misfit dislocations. Microstrains induced by misfit dislocations lead to broadening of XRD profiles. Peak broadening is caused also by small crystallite sizes. These two sources of broadening of XRD profiles were separated by the Williamson-Hall method [9].

Figure 2. SEM micrographs of Mg films deposited on sapphire (0001) substrate at RT (a), at 300°C (b) and at RT and subsequently annealed at 300°C for 1 h (c).

The Mg films deposited on the sapphire substrate exhibit higher broadening of XRD peaks caused by microstrains due to higher density of misfit dislocations compared to the films deposited on the FS substrate. The microstrain and thereby also the density of misfit dislocations exhibit only moderate variations with the deposition conditions. These variations lie within the experimental uncertainties of XRD data.

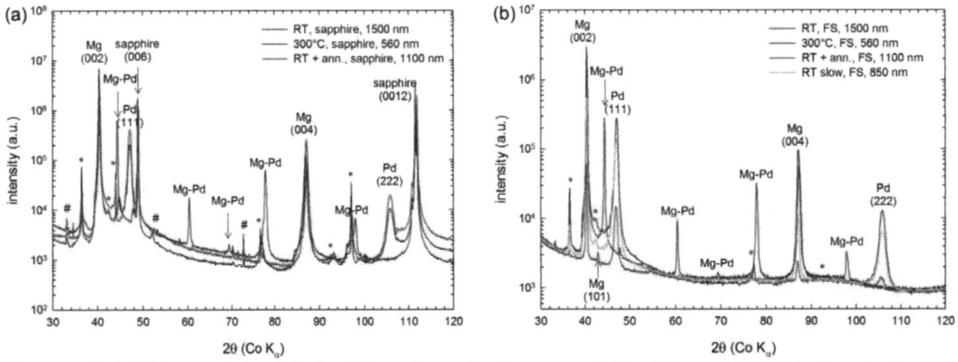

Figure 3. XRD patterns of Mg films deposited on sapphire (a) and FS (b). Symbols "*" and "#" stand for Co K_β reflections and Co lamp artifacts, respectively.

One can see in Figure 3 that annealing of the Mg film with Pd cap at 300°C for 1 h in Ar atmosphere with a pressure of 3 Pa led to disappearance of the both Pd(111) and Pd(222) peaks and five new peaks of an unknown Mg-Pd phase appear. We were not able to identify the Mg-Pd phase formed in the films by comparison with records existing in crystallographical databases. Most probably a metastable phase was formed due to short annealing time, low annealing temperature and an anisotropy of the film. As a rough model we considered a simple hexagonal structure consisting of alternating basal planes filled with Mg and Pd atoms. The *c*-lattice parameter of such structure is 0.475 nm, i.e. the distance between the Mg and the Pd planes ($c/2$) is close to the inter-planar distance of (111) planes in fcc Pd lattice (0.225 nm) and the inter-planar distance of (0002) planes in hcp Mg (0.261 nm). Hence thermally activated diffusion of Pd atoms into Mg layer forms a Pd sublattice. Such structure explains the XRD reflections observed at 44.6°, 68.8° and 97.8° corresponding to (0002), (0003) and (0004) planes, respectively. More precise structure modeling is needed to explain also the the two remaining reflections observed in the spectra.

The dependences of the S parameter on the energy E of incident positrons for two reference bulk Mg samples are plotted in Figure 4a. The red $S(E)$ curve corresponds to a bulk Mg sample annealed at 250°C for 1 h in silicon oil bath. The maximum at ~3 keV is due to positron annihilations in a thin oxide layer formed on the Mg surface during annealing. The sample exhibit a single component positron lifetime spectrum with lifetime 223(1) ps which agrees well with the Mg bulk lifetime obtained from ab-inito theoretical calculations [10]. The mean positron diffusion length $L_+ = 195(8)$ nm obtained from fitting of the *S*(*E*) curve agrees well with the value reported in Ref. [3]. This testifies the annealed bulk Mg contains very low density of defects and virtually all positrons are annihilated in the free state. The blue *S*(*E*) curve in Figure 4a shows data for a bulk Mg sample annealed in a UHV chamber at 250°C and covered by 20 nm Pd over-layer prior to its removal from the chamber to prevent oxidation. The Pd layer was subsequently removed by polishing, hence the the oxide layer was not formed, but polishing induced dislocations into the sample, which led to the shortening of the positron diffusion length down to $L_+ = 64(8)$ nm.

The dependences of the *S* parameter on the positron energy for the Mg films deposited on sapphire substrate are shown in Figure 4b, the data for the bare sapphire substrate are plotted in the figure as well. The *S*(*E*) curves for Mg films exhibit firstly a drop of *S* due to positron annihilations in Pd cap, except of the Mg film #3 (deposited at RT and annealed at 300°C) where Mg-Pd phase was formed. With increasing energy positrons penetrate into the Mg layer and the *S* parameter increases. Due to the different thicknesses of Mg films and varying positron diffusion lengths the *S* parameter approaches maximum at different energies. The maxima of the *S* parameter curves are always higher than the bulk *S* parameter for the annealed bulk Mg. This testifies that Mg films contain defects which trap positrons. When energy of incident positrons is further increased positrons become to penetrate into the sapphire substrate leading to a decrease of *S* which gradually approaches the bulk *S* parameter for the bare sapphire substrate.

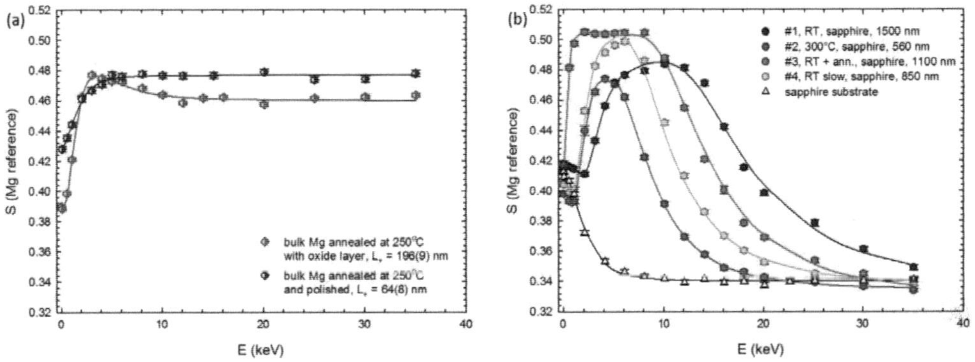

Figure 4. The dependences of the *S* parameter on the energy *E* of incident positrons for two reference bulk Mg samples (a) and Mg films deposited on sapphire substrate (b). Solid lines are model curves calculated by VEPFIT.

The *S*(*E*) curves for Mg films were fitted by VEPFIT assuming 3 layers: (i) Pd cap with thickness of 20 nm, (ii) Mg film with thickness given in Table 1 and (iii) semi-infinite sapphire substrate. In case of the film #3 (deposited at RT and annealed at 300°C) the layer (i) was omitted since Pd diffused into the Mg layer and formed Mg-Pd phase. The model curves calculated by VEPFIT are plotted in Figure 4 by solid lines and are in a good agreement with the experimental points. Each layer is characterized by its positron diffusion length L_+ and the *S* parameter. In order to reduce the number of fitting parameters L_+ and *S* for the substrates were fixed at the values obtained for the bare substrates. All Mg films studied exhibit shorter L_+ and higher *S* than the annealed bulk Mg. It testifies that the Mg films contain open-volume defects which trap positrons. These defects are vacancy-like defects located in a vicinity of grain boundaries and misfit dislocations. According to pulsed low-energy positron beam measurement [4] vacancies and vacancy clusters may be present inside of the grains as well.

Compared to the film deposited at RT the deposition at elevated temperature 300°C leads to coarser grain structure, see Figure 1, i.e. the fraction of grain boundaries and therefore defects which are able to trap positrons is lowered. This leads to the increase of the positron diffusion length and the decrease of the *S* parameter in the Mg layer. On the base of XRD data we suppose the density of misfit dislocations remains approximately unchanged.

sample	thickness	dep. conditions	substrate	L_+	*S*	grain size
#1	1540 nm	RT	sap.	45(2) nm	0.4878(4)	210(6) nm
#1	1540 nm	RT	FS	46(1) nm	0.4843(4)	282(7) nm
#2	560 nm	300°C	sap.	50(5) nm	0.4718(7)	400 (10) nm
#2	560 nm	300°C	FS	n.a.	n.a.	740(30) nm
#3	1100 nm	RT + ann.	sap.	1(5) nm	0.5030(4)	259(9) nm
#3	1100 nm	RT + ann.	FS	n.a.	n.a.	289(7) nm
#4	850 nm	RT slow	sap.	1(3) nm	0.5067(5)	n.a.
#4	850 nm	RT slow	FS	n.a.	n.a.	229(6) nm
–	–	ann.	bulk Mg ann.	196(9) nm	0.4634(5)	–
–	–	ann. + pol.	bulk Mg pol.	64(8) nm	0.4767(8)	–
–	–	–	sap. substr.	46(5) nm	0.3400(2)	–
–	–	–	FS substr.	24(1) nm	0.4367(3)	–

The annealing of the Mg film with Pd cap (sample #3) leads to slightly larger mean grain size and thereby to lower fraction of grain boundaries. However one can clearly see in Figure 4b and Table 1 that the film contains significantly higher amount of defects compared to the films deposited at RT and 300°C. Most likely migration of Pd atoms introduced open-volume lattice defects into the film. However further structural investigations of the Mg-Pd phase are necessary to identify these defects.

At the slow deposition rate Mg atoms have more time to find an equilibrium site compared to the films deposited by fast deposition rate. Thus the interaction between the growing film and the substrate becomes more important. This leads to higher density of defects reflected by the highest value of the *S* parameter and the shortest positron diffusion length among all the Mg films studied.

The film deposited on sapphire substrate at RT exhibits higher value of the *S* parameter than the film deposited on FS under the same conditions. This is in accordance with the author's previous study [5] of the Mg films with thickness ~100 nm.

4. Conclusions

A set of Mg films deposited under various conditions on sapphire and fused silica substrates was prepared by RF magnetron sputtering. The films exhibit nanocrystalline structure, (0001) texture and a negligible in-plane stress. Defect studies by positron annihilation spectroscopy revealed that positrons in the Mg films are trapped in vacancy-like defects at grain boundaries and at misfit dislocations.

Film deposited at 300°C exhibits larger grains and lower concentration of defects than film deposited at RT due to reduced volume fraction of grain boundaries. Annealing of the film covered with Pd at 300°C for 1 h leads to formation of Mg-Pd phase and induces large amount of open-volume defects into the film. Film deposited with slow deposition rate contains highest amount of defects due to the stronger sensitivity to the lattice misfit between the film and the substrate.

Acknowledgement

Financial support by the Czech Science Agency (project P108/12/G043) is gratefully acknowledged.

References

[1] Sandrock G 1999 *J. Alloys Comp.* **293-295** 877

[2] Checchetto R, Bazzanella N, Miotello A, Brusa R S, Zecca A and Mengucci A 2004 *J. Appl. Phys.* **95** 1989

[3] Eijt S W H, Kind R, Singh S, Schut H, Legerstee W J, Hendrikx R W A, Svetchnikov V L, Westerwaal R J and Dam B 2009 *J. Appl. Phys.* **105** 043514

[4] Checchetto R, Bazzanella N, Kale A, Miotello A, Mariazzi S, Brusa R S, Mengucci P, Macchi

C, Somoza A, Egger W and Ravelli L 2011 *Phys. Rev.* B **84** 054115
[5] Hruška P, Čížek J, Anwand W, Bulíř J, Lančok J, Stráská J, Melikhova O and Procházka I 2014 *J. Phys.: Conf. Ser.* **505** 012024
[6] Kirchheim R 1988 *Prog. Mater. Sci.* **32**(4) 261–325
[7] Anwand W, Brauer G, Butterling M, Kissenger H R and Wagner A 2012 *Def. Diff. Forum* **331** 25
[8] van Veen A, Schut H, Clement M, de Nijs J M M, Kruseman A and Ijpma M R 1995 *Appl. Surf. Sci.* **85** 216–224
[9] Williamson G K and Hall W H 1953 *Acta Mater.* **1** 22
[10] Robles J M C, Ogando E and Plazaola F 2007 *J. Phys.: Condens. Matter* **19** 176222

Positron annihilation Doppler broadening measurement for bulk amorphous alloy by using high energy positron generated from LCS gamma-ray at NEW SUBARU

F.Hori[1], Y.Ueno[1], K.Ishii[1], T.Ishiyama[1], A.Iwase[1], S.Miyamoto[2], T.Terasawa[2]

[1] Department of materials science, Osaka Prefecture University, 1-1 Gakuen-cho, Naka-ku, Sakai, Osaka 599-831, Japan

[2] Laboratory of Advanced Science and Technology for Industry, University of Hyogo, 3-2-1 Kouto, Kamigori, Ako-gun, Hyogo 678-1205, Japan

E-mail: horif@mtr.osakafu-u.ac.jp

Abstract. A simple positron annihilation measurement apparatus via pair creation has been developed using high energetic gamma beam generated by laser Compton scattering (LCS) of 1 GeV electrons circulated in a storage ring and laser light with the power more than 1 W at the New SUBARU synchrotron radiation facility, University of Hyogo. This MeV ordered energy changeable positron apparatus is useful to study defects in bulk materials. In this study, the average energy of 8MeV positron was selected by the wavelength of laser light and circulated electron energy in photon factory. As a demonstrate of non-destruction positron measurement by this apparatus, positron annihilation Doppler broadening measurement has performed for bulk size of amorphous and crystal structured Zr based alloys. The larger Doppler broadening S parameter for amorphous alloy than that for crystallized one has been successfully measured.

1. Introduction

Positron annihilation method are very powerful tool for non-destructive test of various materials. This method can give us the information about the part of material where annihilation occurs. Positron implantation depth depends on its energy and target material. Therefore, positrons which have constant a certain energy are the most suitable for measurement. Positron sources we usually use are radio isotope and slow positron beam at some institutes. The slow positrons, which have 10 keV ordered energy, can inject into solids and annihilate at 100 nm depth from the surface. Moreover, the energy profiles of these positrons from radioactive sources have wide distribution. The energetic and monochromated positron beam is the best to get defect information in bulk samples. Electron-positron pair production is one of the simplest methods to obtain energy controlled positrons. Recently, some groups have successfully generated highly energetic positron-electron pairs by pair production from inverse Compton scattering gamma rays at some facilities [2-4]. In the pair creation, the positron energy spectrum is widely spread because of the broadening of the gamma ray (photon) energy. Applying magnetic field, the positrons are separated depending on their own energy. So we can obtain nearly-monochromatic positrons at the position of the target materials. Laser Compton scattering (LCS) is a kind of inverse Compton scattering and utilizes laser lights. We have developed a simple pair creation positron beam apparatus with energy of MeV order by LCS gamma ray from synchrotron radiation electron and incident laser light [1]. Here, we demonstrate an application of this system to the study of the structure's change of internal parts with crystallization of non-destructive bulk amorphous alloy.

2. Experimental procedure

2.1. Apparatus setup

Our positron annihilation apparatus exists at the beam-line (BL01A) of the electron storage ring at the New SUBARU SR facility at LASTI, University of Hyogo. 1.0 and 1.5 GeV electrons circulated in storage ring can use for laser Compton scattering experiment. Also, incident laser can be easy to change, such as Nd, Er, Ti-Safire, and CO_2 in this system and each laser has different wavelength. So by changing the incident laser, the energy of generated positron electron pair be controlled. LCS gamma ray with 16.7 MeV have been generated from 1.0 GeV (200mA) electrons and laser light with a wavelength of 1064 nm from a Nd: YVO_4, which has a power of 4.5W [5, 6, 7]. The generation rate of LCS gamma rays is about 10^5 $MeV^{-1}s^{-1}$ in this setup [8]. The positron production rate of our pair creation system was already measured as 1 positron per 23 photons, and the estimated positron flux is about 3600 s^{-1} [8]. The generated positron and electron pairs are separated by pass through the magnetic field, while the transmitted gamma photons go straight. Figure1 shows the scheme of how to separate positron and electron. The separated positron and electron together with γ-ray distribution at a position x_1+x_2 was measured by using an imaging plate (IP). Figure 2 shows the positron and electron profile on IP after separation. A magnetic field of B = 0.2 T applied behind the Pb target makes the separation. The positrons and electrons in magnetic field are bent into the circle with radius of diameter r_c by Lorentz force $F_L = evB$. From the bending distance from the central point on IP, the radius was able to be calculated. According to the principle of relativity, positron energy E_p is expressed by

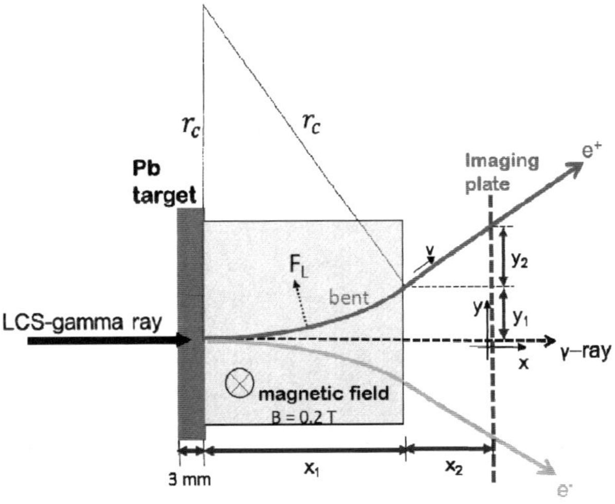

Figure 1. The schematic illustration of pair creation from LCS-gamma ray and separation setup of positron and electron.

$$E_p = \frac{eB}{m'vc^2} = \frac{ceBr_c}{\beta} \quad (1)$$

and r_c is given by

$$r_c = \frac{m'v^2}{F_L} = \frac{m'v}{eB} \quad (2)$$

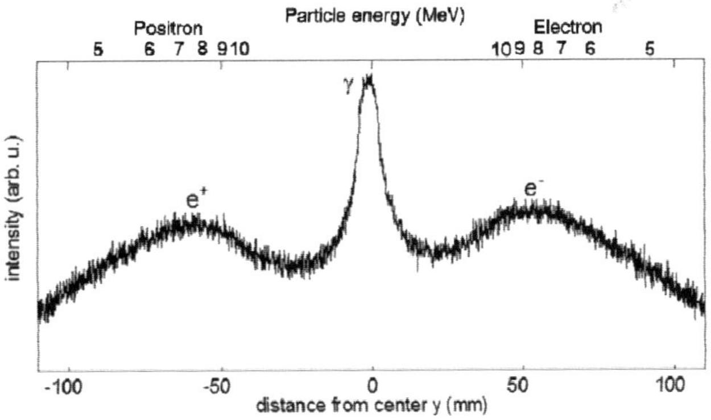

Figure 2. Generated positron and electron profile as a function of distance from center of IP and corresponding positron and electron energy.

where e is the charge of positron, m' is the relativistic mass of positron, v is the velocity of positron, and $\beta = v/c = 0.998$. Distance from center of IP y is the sum of y_1 and y_2 , where $y_1 = r_c(1-\cos\theta)$, $y_2 = x_2\tan\theta$, $\sin\theta \approx x_1/r_c$, x_1=50 mm. When IP is placed at x_2 =120 mm, the calculated y for 8 MeV energy is 56 mm and has a good correlation with the positron peak in this experiment. The separated positrons penetrate into specimen and annihilate with electron in specimen. The positron annihilation gamma-rays are detected at Ge detector placed by the specimen. To reduce the back ground, the detector was covered by 50 mm thick lead blocks and the annihilation gamma rays were restricted by a 30 mm diameter aperture.

2.2. *Bulk sample measurements*

Bulk amorphous alloy samples whose alloy system is $Zr_{55}Cu_{30}Al_{10}Ni_5$ were made by rapidly cooling at a cooling rate of more than 100 K/s. Their shape is a cylinder with 3cm diameter and 4cm length. Figure 3 shows the amorphous and crystalized $Zr_{55}Cu_{30}Al_{10}Ni_5$ alloy. One of them was crystallized by annealing at 500°C for an hour in air. X-ray diffraction measurement using Cu-Kα was carried out for these alloys. Doppler broadening measurement using a Ge detector was performed for these alloys by using LCS-positron apparatus mentioned above. The bulk samples are placed where the peak position of positron profile corresponding to 8 ± 2 MeV. In general, the Makhov profile of positron with positron with energy E is given by

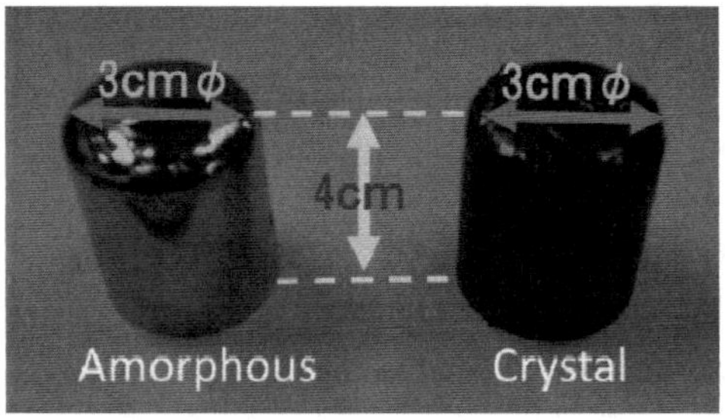

Figure 3. Bulk size amorphous and heat treated crystal alloy.

$$P(z,E) = \frac{mz^{m-1}}{z_0^m}\exp\left[-\left(\frac{z}{z_0}\right)^m\right] \quad (3)$$

with the mean penetration depth given by

$$z_0 = \frac{AE^n}{\rho\Gamma(1+1/m)} \quad (4)$$

where ρ is the mass density of the sample and Γ is gamma function and m , n and A are empirical parameters depending on the material and conditions [9]. As shown in Figure 4, the mean penetration depth of 8 MeV positrons into this sample is estimated to be about 8 mm. In this case, the values used as m , n and A are 2, 1.6 and 4.0, respectively. Considering of energy dispersion and sample diameter, the energy of positron injected into samples corresponds to 6.5-10 MeV. According to these results, the estimated penetrate depth is around 6-12 mm. Doppler broadening spectra consisting of more than $2x10^5$ counts were measured with an accumulation time of 3 to 5 hours.

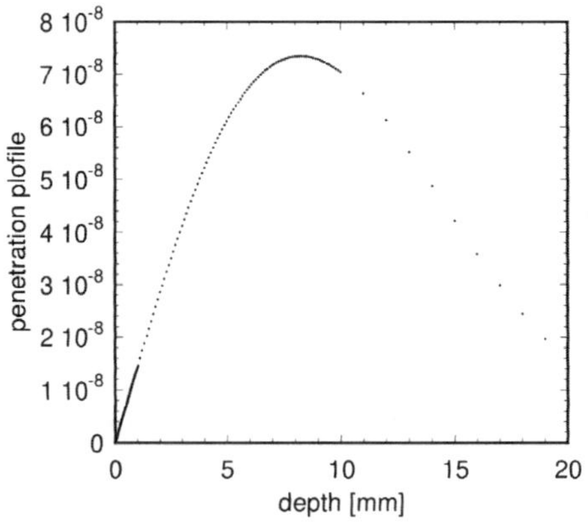

Figure 4. The Makhov profile of positron with 8MeV energy in this sample.

3. Results and discussion

Figure 5 shows XRD spectra of before and after annealed $Zr_{55}Cu_{30}Al_{10}Ni_5$ sample. XRD spectra of amorphous alloy shows halo peak which is peculiar to amorphous. XRD spectra of after annealed sample has several sharp peaks which stem from characteristic X-ray for $Zr_{55}Cu_{30}Al_{10}Ni_5$ crystal alloy [10, 11]. It can be conformed that annealed sample changed from amorphous to crystal structure.

Figure 6 shows the Doppler broadening spectra of before and after annealed sample measured by LCS-gamma positron apparatus. It is well known that the shape of Doppler broadening spectra depends on the alloy system, density and including open volume defects. For $Zr_{50}Cu_{40}Al_{10}$ alloy, it has reported that crystallization from amorphous causes a slight decrease of the intensity of CDB spectrum in electron momentum region near 0 and an increase of density [12]. As can be seen in figure 6, the peak height decreases by crystallization. This density change detected by our new positron instrument is estimated at around 6-12 mm depth. The density of $Zr_{55}Cu_{30}Al_{10}Ni_5$ also increases with crystallization [11], so it can be concluded that this peak height change corresponds to the density change.

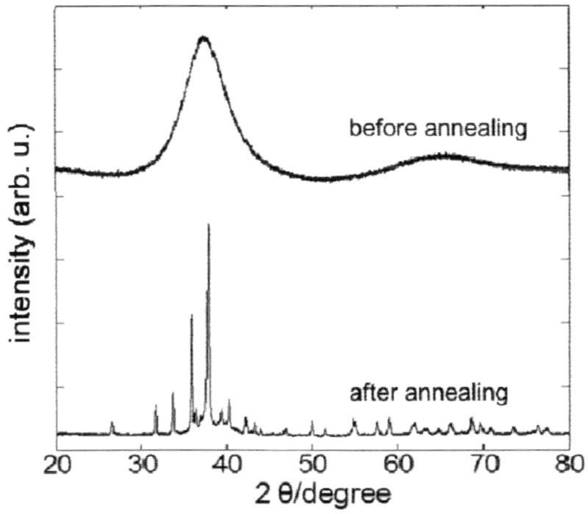

Figure 5. Normalized XRD spectra for $Zr_{55}Cu_{30}Ni_{10}Al_5$ alloys before and after annealing.

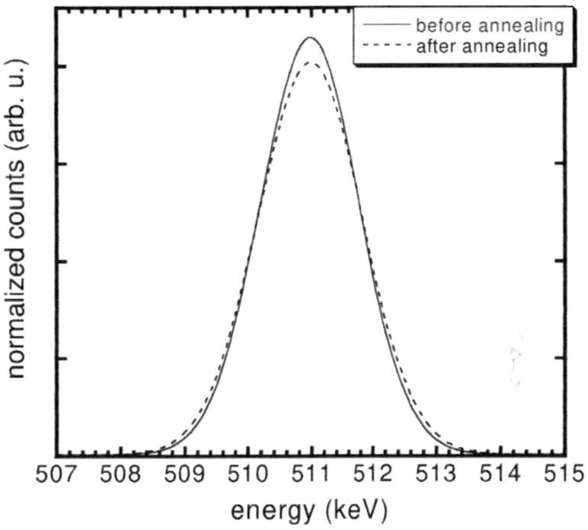

Figure 6. Normalized Doppler broadening spectra for $Zr_{55}Cu_{30}Ni_{10}Al_5$ alloys before and after annealing.

4. Summary

We applied our high energy positron measurement apparatus at New SUBARU for bulk material estimation. The energy of positron created by pair creation from LCS gamma-ray was confirmed 8 MeV by the measurement on IP. Using the apparatus of high energy positron beam, we have successfully measured bulk samples in mm order depth without destruction. We found that the mm ordered internal structure of bulk amorphous alloy changed by annealing.

References

[1] Hori F, Ishii K, Ishiyama T, Iwase A, Miyamoto S, and Terasawa M 2014 *Jpn. J. Appl. Phys. Conf. Proc.* **2** 011301

[2] Dobayashi K, Hirose T, Kumita T, Kurihara Y, Muto T, Omori T, Okugi T, Sugiyama K and Urakawa J 1999 *Nucl. Inst. And Meth.* A **437** 169

[3] Hugenschmidt C, Schreckenbach K, Habs D and Thirolf P.G 2012 *Appl. Phys.* B **106** 241

[4] Hirade T, Toyokawa H, Ohdaira T, Suzuki R and Ohgaki H 2004 *Mater. Sci. Forum* **445-446** 474

[5] Li D, Imasaki K, Miyamoto S, Horikawa K, Amano S and Mochizuki T 2009 *Appl. Phys. Let.* **94** 091112

[6] Horikawa K, Miyamoto S, Amano S and Mochizuki T 2010 *Nucl. Inst. and Meth.* A **618** 209.

[7] Amano S, Hoirkawa K, Ishihara K, Miyamoto S, Hayakawa T, Shimizu T and Mochizuki T 2009 *Nucl. Inst. and Meth.* A **602** 337

[8] Horikawa K, Miyamoto S, Amano S, Li D, Imasaki K and Mochizuki T 2010 *IEEJ Trans. EIS* 130 **10** 1784

Journal of Physics: Conference Series **674** (2016) 012025 doi:10.1088/1742-6596/674/1/012025

[9] Schultz P.J and Lynn K.G. 1988 *Rev. Mod. Phys.* **60** 701
[10] Kondoh K, Kawabata K, Serikawa T and Kimura H 2008 *Trans. of JWRI.* Vol. **37** 312057
[11] Inoue A, Negishi T, Kimura H.M, Zhang T and Yavari A.R 1998 *Mater. Trans, JIM* Vol. **39** No. 2 318
[12] Yano T, Yorikado Y, Akeno Y, Hori F, Yokoyama Y, Iwase A, Inoue A and Konno T.J 2005 *Materials Transactions*, Vol. **46** 2886

Institute of Physics
Dirac House, Temple Back
Bristol BS1 6BE UK

ISSN: 1742-6588
ISBN 978-1-5108-1965-8

CURRAN ASSOCIATES INC.
proceedings.com